RAINFORESTS

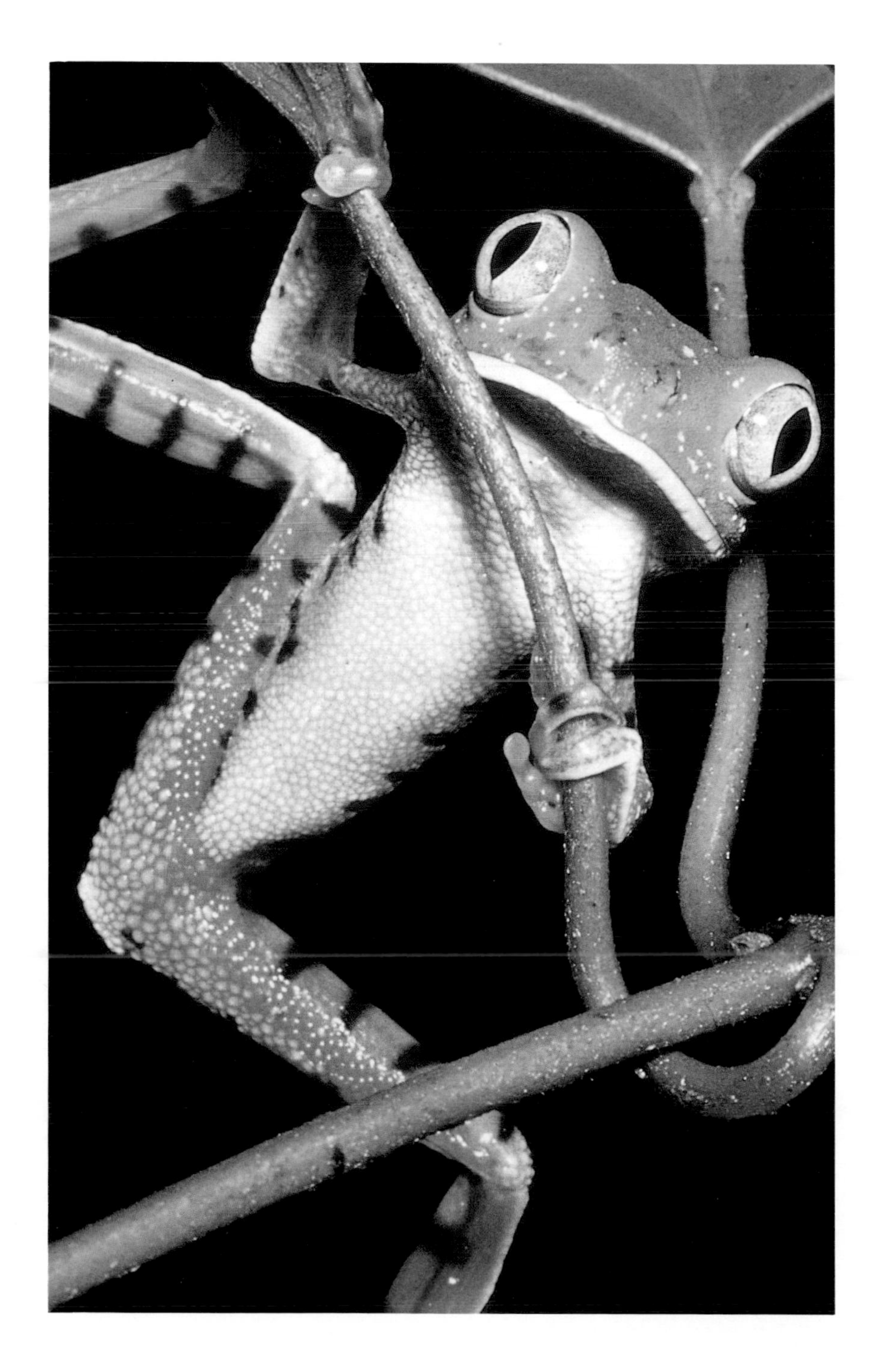

RAINFORESTS

The Illustrated Library of the Earth

CONSULTING EDITOR

NORMAN MYERS, Ph.D.

RODALE PRESS
EMMAUS, PENNSYLVANIA

Published 1993 by Rodale Press, Inc.
33 East Minor Street, Emmaus, PA 18098, USA

By arrangement with Weldon Owen
Conceived and produced by Weldon Owen Pty Limited
43 Victoria Street, McMahons Point, NSW, 2060, Australia
Fax (61 2) 929 8352
A member of the Weldon International Group of Companies
Sydney • San Francisco • London

CHAIRMAN: Kevin Weldon
PRESIDENT: John Owen
GENERAL MANAGER: Stuart Laurence
PUBLISHERS: Sheena Coupe, Alison Pressley
SERIES COORDINATOR: Tracy Tucker
ASSISTANT EDITOR: Veronica Hilton
COPY EDITORS: Carl Harrison-Ford, Tracy Tucker
DESIGNER: Toni Hope-Caten
DESIGN CONCEPT: Andi Cole, Andi Cole Design
COMPUTER LAYOUT: Paul Geros, Veronica Hilton, Steven Lloyd
PICTURE RESEARCH: Jenny Mills, Brigitte Zinsinger
ILLUSTRATION RESEARCH: Joanna Collard
CAPTIONS: Terence Lindsey
INDEX: Diane Harriman
MAJOR ILLUSTRATIONS: Jon Gittoes
OTHER ILLUSTRATIONS AND MAPS: Mike Gorman
COEDITIONS DIRECTOR: Derek Barton
PRODUCTION DIRECTOR: Mick Bagnato
PRODUCTION COORDINATOR: Simone Perryman

Library of Congress Cataloging-in-Publication Data

Rainforests/consulting editor, Norman Myers.
p. cm.—(The Illustrated library of the earth)
"A Weldon Owen production"—Verso t.p.
Includes index.
ISBN 0-87596-597-0 hardcover
1. Rain forests—Tropics. 2. Human ecology—Tropics. 3. Rain forest ecology. 4. Man—Influence on nature—Tropics. 5. Deforestation—Tropics. 6. Conservation of natural resources—Tropics. I. Myers, Norman. II. Weldon Owen Pty Limited. III. Series.
GF54.5.R35 1993
304.2'0915'2—dc20 92-46131
CIP

If you have any questions or comments concerning this book, please write to:
Rodale Press
Book Readers' Service
33 East Minor Street
Emmaus, PA 18098

Production by Mandarin Offset, Hong Kong
Printed in China

Distributed in the book trade by St. Martin's Press

10 9 8 7 6 5 4 3 2 1

A WELDON OWEN PRODUCTION

JACKET: *A view of Lamington National Park, Queensland, Australia. Photo by Grenville Turner/Wildlight Photo Agency.* ENDPAPERS: *A three-toed sloth in the leafy canopy of an Amazonian forest. Photo by M.P.L. Fogden/Bruce Coleman Limited.* JACKET INSET & PAGE 1: *A barred leaf-frog. Photo by Michael & Patricia Fogden.* PAGE 2: *Raggiana's bird of paradise, New Guinea. Photo by D. Parer & E. Parer Cook/AUSCAPE.* PAGE 3: *Raoni, chief of the Kayapó tribe of Amazonia. Photo by Bill Leimbach.* PAGES 4–5: *Squirrel monkeys are common inhabitants of the forests of Latin America.* PAGES 6–7: *Emerald tree-boas give birth to live young, which often vary markedly in color.* PAGE 8: *A barred leaf-frog perches on a heliconia flower in the Amazon rainforest. Photo by Michael & Patricia Fogden.* PAGES 10–11: *A gathering of Brazilian Kayapó.* PAGES 12–13: *Australian green tree ants.* PAGES 80–81: *New Guinea. highlanders.* PAGES 130–131: *A ravaged landscape: Madagascar.*

Jany Sauvanet/AUSCAPE

Contributors

DR. PETER BERNHARDT
Associate Professor of Botany, St. Louis University, Missouri, USA, and Research Botanist, Royal Botanic Gardens, Sydney, Australia

DR. FRANCIS H.J. CROME
Principal Research Scientist, Division of Wildlife and Ecology, Commonwealth Scientific and Industrial Research Organization (CSIRO), Australia

DR. JAMES A. DUKE
Economic Botanist, National Germplasm Resources Laboratory, United States Department of Agriculture, Maryland, USA

DR. TERRY L. ERWIN
Department of Entomology, National Museum of Natural History, Smithsonian Institution, Washington DC, USA

DR. ADRIAN FORSYTH
Indonesia Program Director, Conservation International, Washington DC, USA

DR. CARL F. JORDAN
Senior Ecologist, Institute of Ecology, University of Georgia, Athens, USA

JUSTIN KENRICK
Department of Social Anthropology, University of Edinburgh, UK

DR. AILA KETO
Rainforest Conservation Society, Queensland, Australia

DR. THEODORE MacDONALD, JR.
Projects Director, Cultural Survival, Inc., and Research Associate, Department of Anthropology, Harvard University, Massachusetts, USA

JEFFREY A. McNEELY
Chief Conservation Officer, International Union for Conservation of Nature and Natural Resources (IUCN), Switzerland

ANDREW MITCHELL
Deputy Director, Earthwatch Europe, Oxford, UK

PROFESSOR NORMAN MYERS
Independent Consultant in Environment and Development, Oxford, UK

PROFESSOR FRANCIS E. PUTZ
Associate Professor, Department of Botany, University of Florida, Gainesville, USA

DR. MICHAEL H. ROBINSON
Director, National Zoological Park, Smithsonian Institution, Washington DC, USA

PAUL SPENCER WACHTEL
Head, Creative Services, World Wide Fund for Nature International, Switzerland

DR. MICHAEL WILLIAMS
Reader, School of Geography, University of Oxford, UK

Jany Sauvanet/AUSCAPE

Contents

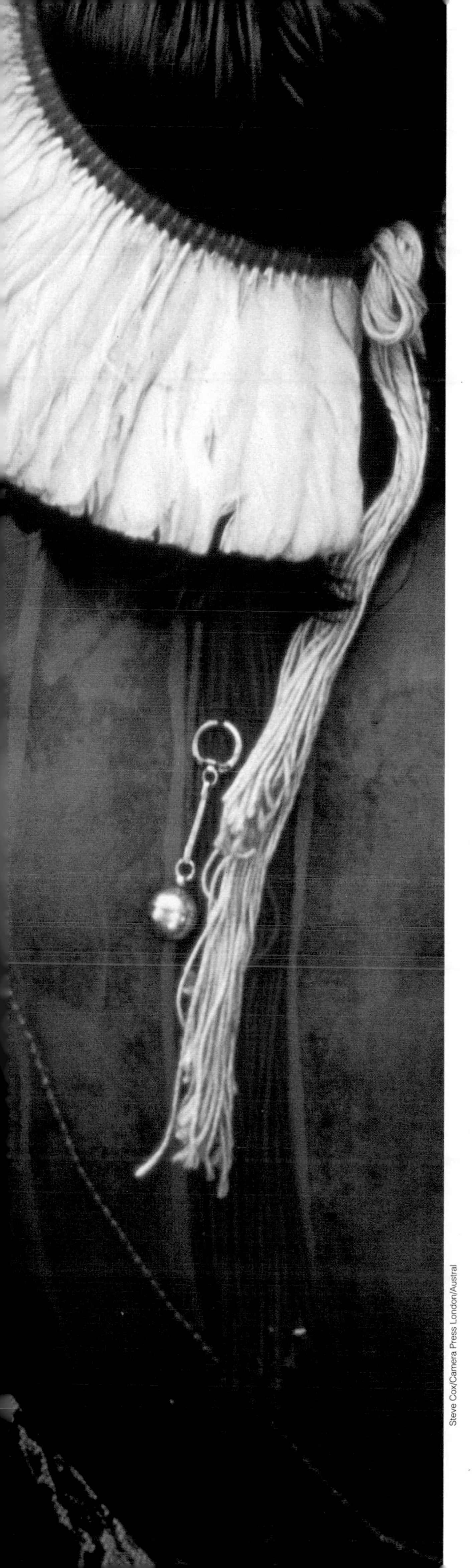

Steve Cox/Camera Press London/Austral

Foreword

The addition of some three billion people to a record human population of 5.4 billion in 1992 will take place over the next 30 to 40 years. Almost all of the growth will take place in developing countries, where over three-quarters of the world's people live now. More people will live in Mexico City by the end of the 1990s than inhabited all of the cities of the world in 1800, and urban dwellers seem to find it more easy than their rural counterparts to forget their complete and utter dependency on the natural world.

The present population of the world, which has more than doubled in the past 40 years, is changing the characteristics of the atmosphere and of regional and global climates, wasting topsoil on the world's agricultural lands at an incredible rate (more than a fifth lost since 1950), and destroying the world's species of plants and animals at a rate 1,000 to 10,000 times that characteristic of the past 65 million years of Earth history. One in five of the world's people is living in extreme poverty, and one in ten is malnourished.

Economic principles that may have served the world well in a period of expansion continue to treat it and all of its resources, in the words of economist Herman Daly, as if it were "a business in the course of liquidation". As our numbers and the complexity of our civilizations have grown, it has become increasingly difficult to deal with our rich, complex, and potentially sustainable world in a manner that shows any regard for those who will follow us. All over the world, potentially renewable resources such as the tropical forests are being destroyed for one-time profit, without any regard for the future, while poor people, largely ignored by their governments and by the world at large, are ravaging the forests and exhausting other resources in an endless quest for survival.

This book was prepared by world-class experts. Its lucid writing, sensitivity to the people who interact with the tropical forests, and excellent illustrations provide a sound, factual basis for understanding tropical forests. Read and understood, it will make a major contribution to our ability to deal with tropical forests intelligently, and thus to pass on our world in the best possible condition to our children and grandchildren.

PETER H. RAVEN

Director, Missouri Botanical Garden, St Louis, Missouri

PART ONE

LAYERS OF TEEMING LIFE

Diversity is the hallmark of the tropical rainforest—no habitat rivals it for sheer numbers of species. And because so many species of plants and animals interact, evolution has had an opportunity to fine-tune their adaptations.

Densey Clyne/Mantis Wildlife

1 The World's Tropical Forests

MICHAEL H. ROBINSON

The rainforest, once a girdle of green around the globe, has been reduced by more than half since the beginning of this century. In the past 10 years alone the rate of its destruction has increased by 90 percent. In what is now the USA the first European settlers found forests covering nearly 170 million hectares (420 million acres) of which only about 10 million hectares (25 million acres) remain. Here is environmental perturbation on a scale that makes the word seem trivial. What are tropical forests, where are they, and why do they matter?

Untold Variety

Tropical forests are not simple to define. There are many kinds ranging from very wet to very dry. They may be of different composition and different architecture, and they range from monumental in the lowlands to dwarf or fairy forests on high mountains. But the majority share one outstanding feature: they are teeming with life. Certainly they are the treasure-houses of biology. Fort Knox, the Prado, the Louvre, the Hermitage, the British Museum—think of any place that represents a unique concentration of cultural or material wealth and then apply that image to the tropical forests. In mere numbers of species they are overwhelmingly the center of life on Earth. In fact no accumulation of human artifacts even comes close to the number of different "objects" in the rainforest. No library anywhere can match the information stored in the chromosomes of rainforest life.

Stephen Krasemann/NHPA

In temperate regions, orchids are very often terrestrial in habit but in tropical rainforests they are commonly arboreal, and sometimes restricted to the sunlit upper canopy. This is Sobralia powoli *of Central America.*

Opposite. *Ten meters (30 feet) above the ground in a Costa Rican rainforest. In many rainforests, a distinct sub-canopy zone exists, made up not only of young forest giants growing through, but also of tree species that have evolved to suit moderate light levels between the extremes of constant gloom below and sunlit glare above. Palms are often prominent in such communities.*

This is no ranting hyperbole. Recent estimates of animal species in the rainforest are between 5 and 30 million, possibly many more. It is surprising to most people, and even to non-biological scientists, that at the end of the twentieth century there should still be such a scatter of estimates. But the reason for such imprecision is the unknown size of the undiscovered and undescribed fraction that resides, overwhelmingly, in the tropics. It consists of those small creatures, mainly animals without backbones, that are the "movers and shakers" of the system. Insects are the most numerous creatures on Earth in biomass (total weight of organisms in a given area) and species; they blossom mightily in the tropics, dwarfing their variety in all other regions into insignificance. Recent estimates based on sampling of rainforest canopies suggest that there may be as many as 30 million species of insects, instead of the one to two million of earlier estimates. Even if this figure proves two or three times too high it still means that the tropics contain more than 90 percent of all species.

The difficulty of counting insect species compared with vertebrates is obvious, and it is also greatly compounded by the vastness, impenetrability, and inaccessibility of the tropical forests. A further exacerbating factor is that while history has determined that the majority of the world's biologists live in the developed temperate regions, evolution has determined that the majority of their subjects live in the tropics. It is scarcely surprising we don't even have consensus on the number of species.

Nor are the riches of the tropical forests confined to numbers of species and biomass. Because so many species of plants and animals in the tropical forests interact for more of each year than anywhere else on Earth, evolution has had an opportunity to fine-tune adaptations. The tropics are the cockpit of evolution where predators and prey, parasites and host, competitors for resources and rivals for mates have battled mightily. They are the center of an evolutionary arms race that has produced a dazzling range of weapons, defenses, specializations, and adaptations of great complexity. This is where most of nature's most exciting and bizarre "inventions" originated.

For instance, predator-to-prey interactions are played out in the tropics with specialized senses that

FROM THE TREETOPS TO THE FOREST FLOOR

Elements of different rainforest types are brought together in this South American panorama illustrating many characteristic rainforest features: tall trees with straight trunks and high crowns, buttress roots, vines and other creepers, bromeliads and other epiphytes, and forest floor fungi. Some of the creatures that inhabit the South American rainforests are listed below.

KEY TO ANIMALS

1 *harpy eagle*
2 *red howler monkeys*
3 *silky (pygmy) anteater*
4 *kinkajou*
5 *Linne's two-toed sloth*
6 *Geoffroy's spider monkey*
7 *tree porcupine*
8 *Toco toucans*
9 *zebra butterfly*
10 *bronzy hermit hummingbird*
11 *emerald tree boa*
12 *northern tamandua*
13 *tree frog*
14 *scarlet macaws*
15 *jaguar*
16 *anole lizard*
17 *ringtailed coati*
18 *ocelot*
19 *quetzal*
20 *Brazilian tapir*
21 *Pallas' long-tongued bat*
22 *cock of the rock*
23 *anaconda*
24 *black caiman*
25 *bushmaster snake*
26 *capybara*
27 *reticulated poison-arrow frog*
28 *giant armadillo*
29 *bird-eating spider*
30 *hoatzin*

High above the ground, a sea of foliage seems to extend almost unbroken from one forest boundary to the other. This world of leaves is home to many species of birds, mammals and reptiles, as well as a multitude of insect species so far almost entirely unexplored.

Though lacking the bright sunshine of the upper canopy, the middle forest levels are sheltered from extremes of sun, wind, and rain by the blanket of leaves above. Mammals are fewer at this level but many birds live only here, and the bark of trunks and lower limbs of trees harbors a wealth of insects and other small life forms.

Entirely protected from the elements, the ground is the most stable forest environment. Few winds penetrate, and temperature, humidity, and ambient light levels almost never change. Rainforests often stand on thin, poor soils; nutrients cycle so rapidly that there is little delay between their abandonment by one life form and absorption by another. Wide buttresses, characteristic of many rainforest trees, often betray the shallowness of root systems below the soil.

almost defy the imagination. It is now almost a commonplace wonder that sonar in bats allows them to find insects in the dark, but tropical fishing bats also use sonar to find fishes and have a gaff-like toe with which to catch them. On the other side of the arms race, some of the insects hunted by bats have evolved ears to pick up the bats' supersonic calls, and the ability to take complex evasive action. Fishes in turbid forest rivers have electrosenses to find their way around, interact, and survive.

The strategies of hide-and-seek have also reached an unsurpassed state in the tropics. Camouflage and concealing mimicries abound. Indeed, some are so "perfect" that more sceptical biologists question their function, claiming they seem better than necessary. But such adaptations speak to the extraordinary discriminatory powers of predators, which are themselves a response to the progress in defense.

Similarly, specialized feeding habits allow animals to outcompete rivals by utilizing resources not available to generalists. This is an evolutionary force that can drive parts of the system to narrower and narrower niches. Thus there are frog-eating bats, blood-sucking moths, mites living on the jaws of army ants, arachnids that hitch flight-time on the bodies of beetles, and flies that share the digested food of spiders, to name but a few examples of biological odd-couplings. Associations between plants and animals also reach strange heights of complexity. Ants are a crowning example of such diversity; they may harvest seeds, grow funguses on leaf-compost, and defend trees against pests and competitors. Plants defend themselves in the great game too. They have chemical defenses against funguses, bacteria, viruses, insects, vertebrate grazers, and browsers. No wonder the tropics have been described as the greatest pharmaceutical factory on Earth. No wonder it is in the tropical forests that we are finding new products and new medicines.

Tropical Rain and Sun

Usually when people speak of tropical forests, or jungles, they are referring to rainforests or moist

Below. *Fruits known as "little lemons" growing on a tree in a Costa Rican rainforest. One fruit has split open, revealing contrasting seeds upon a fleshy pink bed. These fruits are probably attractive to birds, which largely rely on sight for finding food.*

Center. *A view over the Karawari River area, Papua New Guinea. The canopies of tropical forests like this are the biological powerhouses of the planet, the global airconditioners where uncountable numbers of leaves control humidity, soak up carbon, and exhale oxygen.*

K. G. Preston-Mafham/Premaphotos Wildlife

Michael Yamashita

forests. Of course, tropical forests are not all rainforests, since any forest occurring within the limits of the tropics of Capricorn and Cancer deserves the adjective "tropical". At sea level, tropical areas have very small daily or yearly temperature fluctuations. This may be as little as 6°C (11°F) near the equator and up to 12°C (22°F) near the tropics. Twenty degrees Celsius (36°F) seems to be an arbitrary, but generally valid and acceptable figure for the lower temperature boundary of tropical rainforest.

But "rain" in the word "rainforest" is crucially important. Within the tropics there are dry forests, often labelled xeric, as well as montane forests in which the effect of high rainfall is offset by the extent of water-loss due to exposure. These are not rainforests, which are best distinguished from other tropical forests by both the amount of rain they receive and its annual distribution.

Rain in rainforests may be so astonishing that travelers, scientists, and explorers find themselves totally unprepared for its quantity or intensity when they first experience a deluge. To stand in a tropical forest and hear the deafening roar of a torrential rainfall, in the New World perhaps accompanied by the "rainsong" of howler monkeys, is an experience that few temperate-region humans can comprehend. In some tropical forests annual rainfall may reach the remarkable volume of more than 11,000 millimeters (430 inches), and many areas exceed 2,500 millimeters (100 inches). In Central Panama, for instance, rainfall in the Pacific coast forests may average 1,800 millimeters (71 inches) a year, while that in Atlantic-side forests reaches 3,277 millimeters (129 inches). The scientific research station at Barro Colorado Island in Panama has, in the past 50 years, had an average annual rainfall of 2,616 millimeters (103 inches).

Plants and animals have adapted to this intensity in a wide variety of ways. Leaves have complex drip tips that funnel the rain off their surfaces and prevent their leaf-cells from being inundated. Animals can die of thermal stress if they get wet, so they have sheltering behaviors. Even the omnipresent web spiders take

Slow movement, a low metabolism, and dense fur characterize many leaf-eating arboreal rainforest mammals, reflecting the relatively low food value of such a diet. Sloths are an especially notable example, but other unrelated species, like this spotted cuscus Phalanger maculatus *of tropical Australia, show more than a hint of this tendency.*

Pavel German

their webs down before the weight of water collapses them and the silk is irretrievably lost.

Rain is a conspicuous drama in rainforests, where it produces a uniformly high humidity. Significantly, in rainforest areas of great extent such as the Amazon basin, evapotranspiration (water-loss from the leaves of the canopy) leads to cloud-production and initiates a system of rain-making elsewhere in the area. Rainforests both depend on climate and themselves affect and help maintain it.

The volume of rain determines the rainforest, but not in a simple way. Close to the equator, heavy rain may fall more or less constantly—not always, but in most places. This produces evergreen forest in which there is no marked autumn and trees do not necessarily shed their leaves at a particular time. Most trees do shed their leaves, but mainly for biological rather than climatological reasons. Beyond the equatorial region rainfall is seasonal and often depends on the seasonality of the monsoon winds that bring the monsoon rains. The weather patterns are complex and varied, and in some regions there may be two monsoon periods in a year and more than one dry season. In Sri Lanka, for example, some forests are rained upon by southwest and southeast monsoons at different times of the year, with marvellous effects. When a rainforest is seasonal there is a marked biological difference between the wet and dry seasons. Trees flower in all seasons, fruits and leaves fall in all seasons, life propagates in all seasons, but the dry season may be the preponderant period of leaf-fall and flowering. The start of the wet season may be the peak time for birds to rear their young. All this is a fairly simple rhythm, but seasonalities may involve some complexities that are not obvious at first.

One of these involves sunlight and clear skies. The dry season is usually so cloudless and sunny that plants and animals have to deal with the problems of exposure to the sun's rays rather than temperate winter's problems of insulation. Some regions have constantly cloudy wet seasons, others have intermittently sunny parts of their wet seasons; the result is not one kind of seasonal rainforest but a complexity of forms. This has inevitably led to a complexity of systems of scientific classifications for tropical forests. These systems are based on different criteria according to the priorities of the inventors and can be correlated. But except for the specialist, they are easy to ignore. We cannot here delve into hair-splitting. A tropical forest that is toweringly splendid, outstandingly diverse, and very humid for most of the year is a rainforest for most people, a jungle for the remainder. This must be a sufficient definition for this book.

The leaves of many canopy tree species are regular in shape and form. The pointed tips, or so-called "drip tips", and smooth surfaces of many leaves allow them to shed water quickly after rain. This form of foliage seems to discourage the growth of epiphylls on leaves, thus avoiding shading of the photosynthetic cells.

Michael & Patricia Fogden

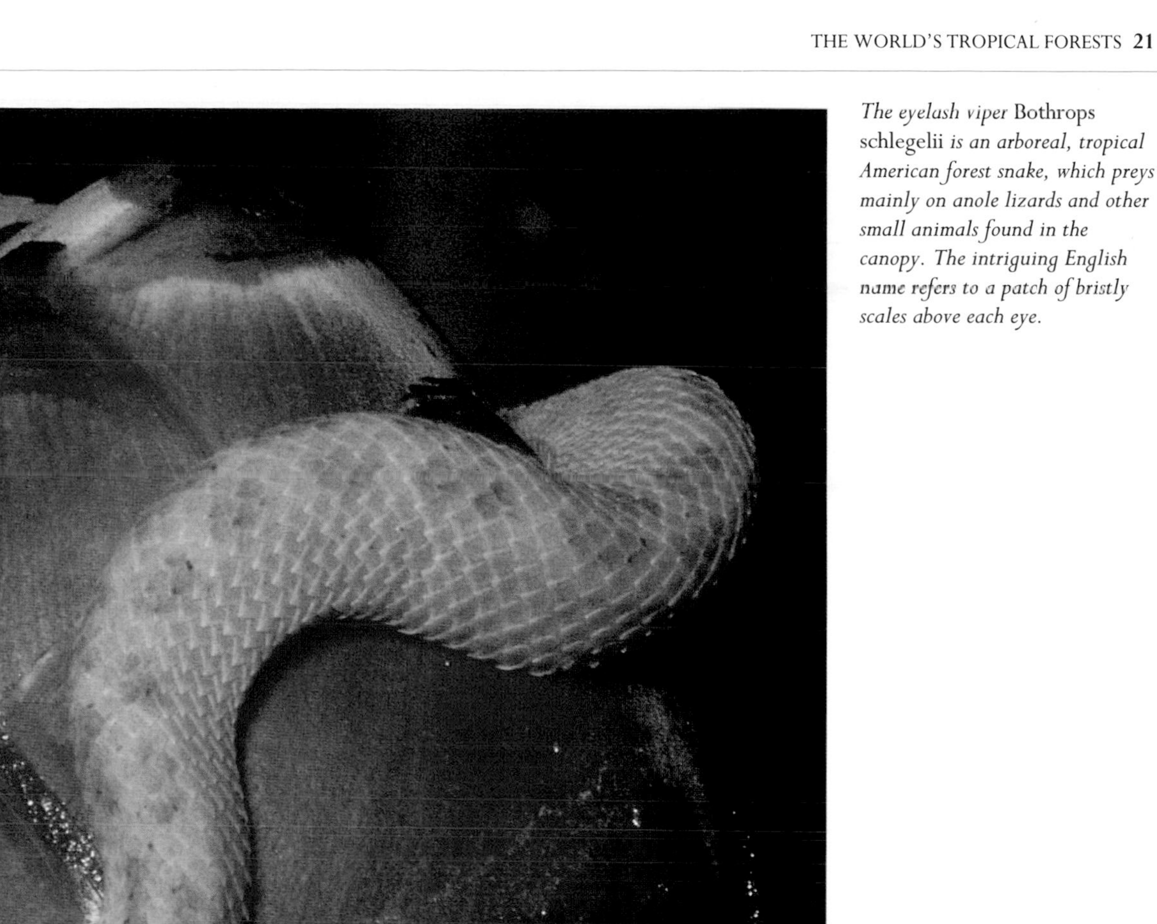

The eyelash viper Bothrops schlegelii *is an arboreal, tropical American forest snake, which preys mainly on anole lizards and other small animals found in the canopy. The intriguing English name refers to a patch of bristly scales above each eye.*

TROPICAL RAINFOREST DISTRIBUTION

NEOTROPICAL REALM		AFROTROPICAL REALM			
Belize	Guatemala	Angola	Equatorial Guinea	Madagascar	Sierra Leone
Bolivia	Guyana	Benin	Ethiopia	Malawi	Somalia
Brazil	Honduras	Burundi	Gabon	Mauritius	Sudan
Caribbean nations	Mexico	Cameroon	Gambia	Mozambique	Tanzania
Colombia	Nicaragua	Central African Rep.	Ghana	Nigeria	Togo
Costa Rica	Panama	Comoros	Guinea	Rwanda	Uganda
Ecuador	Peru	Congo	Guinea-Bissau	São Tomé & Príncipe	Zaïre
El Salvador	Surinam	Ivory Coast	Kenya	Senegal	Zambia
French Guiana	Venezuela	Djibouti	Liberia	Seychelles	Zimbabwe

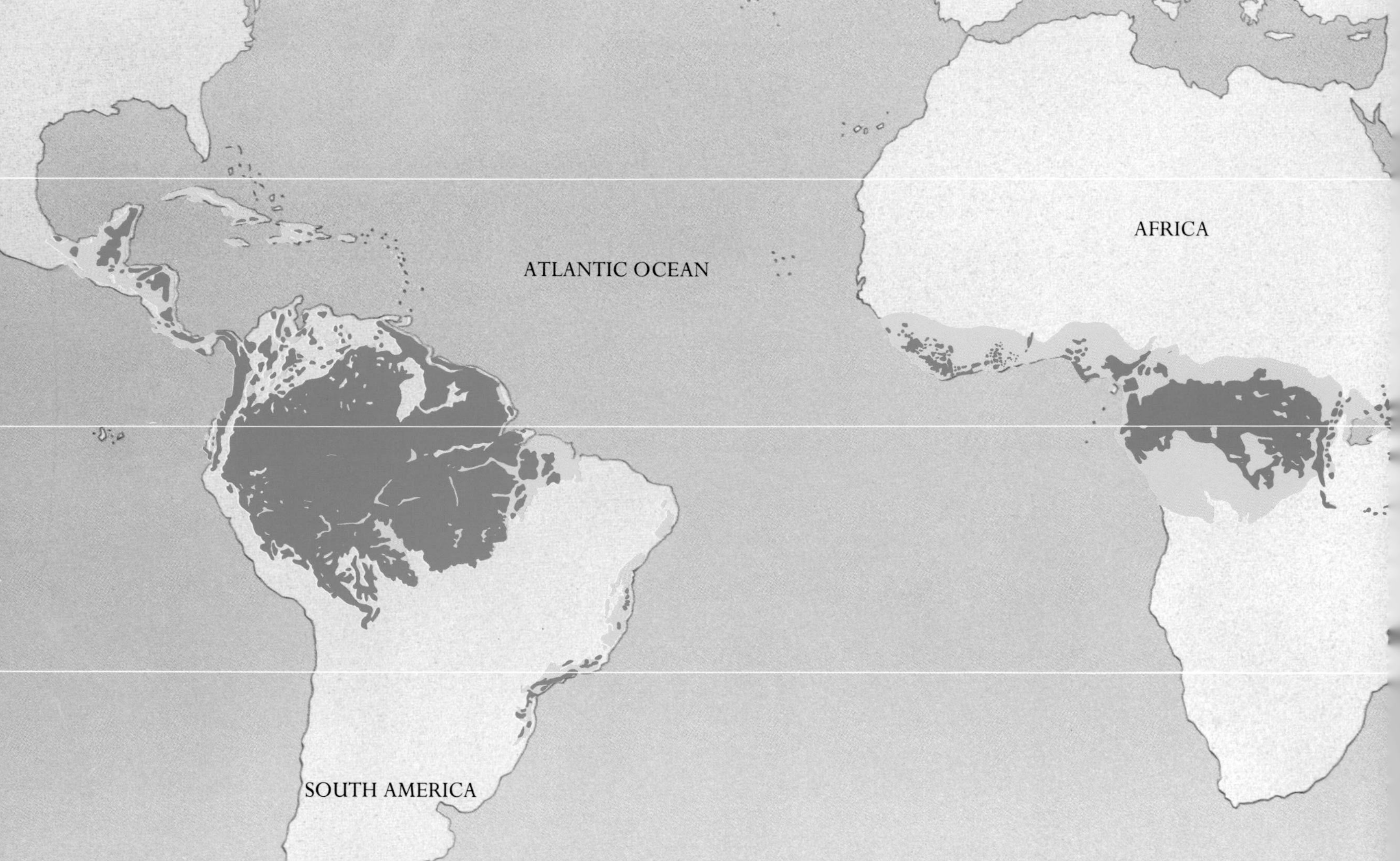

The Realms of the Rainforest

The great rainforests can be clearly located in four biogeographic realms. These are: the Afrotropical realm (sub-Saharan Africa and Madagascar); the Neotropical realm (South America, Caribbean, and lowland Central America); the Indomalayan realm (India, Southeast Asia, Philippines, and most of Indonesia); and the Australian realm (Australia and the island of New Guinea). Of these the Neotropical rainforests, comprising the Amazonian complex as well as Central America and other offshoots west of the Andes, are the largest. By some estimates, they contain several times more moist tropical forest than all other realms combined, although this depends on how one classifies the forest and the regions. Whatever criteria are used it remains the region with the most forest still intact. Some rainforests are still relatively undisturbed,

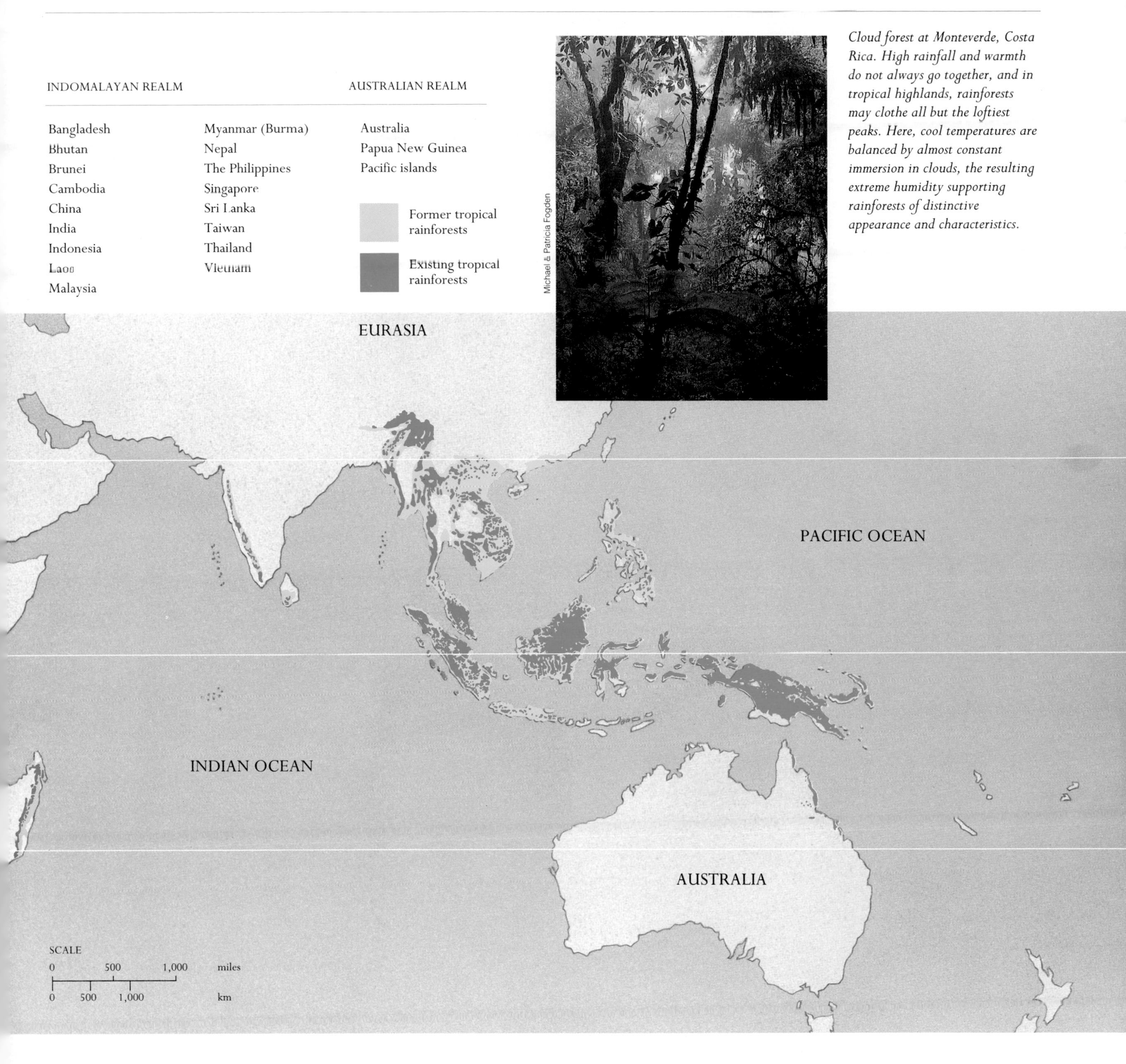

Cloud forest at Monteverde, Costa Rica. High rainfall and warmth do not always go together, and in tropical highlands, rainforests may clothe all but the loftiest peaks. Here, cool temperatures are balanced by almost constant immersion in clouds, the resulting extreme humidity supporting rainforests of distinctive appearance and characteristics.

whereas others have been extensively modified. With their low population density, Papua New Guinea and Irian Jaya (which together form the island of New Guinea) comprise one region where much of the forest must be close to its condition at the beginning of this century. In fact even the most pessimistic assessment suggests that 85 percent of the original forested area on that large island is still forested. The Ivory Coast and Thailand, on the other hand, retain only 10 and 17 percent respectively of their original forest cover. In terms of political geography it is striking to note that in 1989 Brazil, Indonesia, and Zaïre together contained more than 52 percent of the world's remaining tropical forests. Add to these Burma, Gabon, Colombia, Peru, Venezuela, the Guianas, and New Guinea and they account for 82 percent of all the tropical forests. The small remainder is split between some 57 or so other countries.

M.P.L. Fogden/Bruce Coleman Limited

Michael & Patricia Fogden

Top. *Rainforest fruits are often dispersed by birds, which digest the pulp but excrete the entire seed. This is the fruit of an African species of* Connarus, *a genus of shrubs and trees widespread in tropical forests around the world. Some species, such as the zebra woods, are important timber resources.*

Bottom. *Vines climb rainforest trees in one of four ways: scramblers use barbs and hooks; clingers send out aerial roots; twiners coil themselves around the trunk; and tendril climbers send out coiled, sensitive tendrils that take full advantage of even the flimsiest support—like this Mexican cucurbit vine* Momordica *(a close relative of pumpkins and melons).*

The Struggle for Light

Diagrams that show the vertical profile of the typical rainforest are useful if viewed in perspective, but diagrams suggesting that the rainforest is neatly layered may raise expectations that reality does not match. The problem arises quite simply from the fact that, although plotting the distribution of trees and woody shrubs that constitute a forest may reveal a distributional or statistical layering, visitors to the forests do not usually perceive any such clear stratification. The aspect from within the forest is confusingly complex, and quite literally one cannot usually see the "woods" (layers, strata, stories, or tiers) for the trees. The argument between tropical botanists and foresters about stratification continues to be heated. The problem is that the forest is a dynamic system in which ageing and renewal are actively occurring processes somewhere within the system. Nevertheless, detailed analysis of many forests reveals—at some points, and at some times—a stratification that is not obvious on casual observation.

Most mature rainforest has a closed canopy. This occurs where the edge of the leafy-green crown of one tree closely abuts the edge of its neighbor's. These boundaries fit together in irregular mosaics. The result is not an even and uniform meadow or pasture of leaves, high above the ground, but a complex and varied landscape. In relatively flat lowland areas it forms something resembling the surface of plastic bubblewrap, continuous but dimpled and domed. From a closed canopy of this kind a number of giant trees may thrust upwards like umbrellas in a crowd. These are called emergents, and different genera and species tend towards emergent growth in different parts of the tropics.

The canopy and its emergents form the area of the forest that receives the overwhelming majority of solar energy. It is the region of primary productivity, where the photosynthetic machinery of the leaves produces the food resources that support the distribution and supply services of the rest of the tree, and its secondary machinery. The huge production apparatus in the canopy also supports, directly or indirectly, most of the animal wealth of the rainforest. In an ecosystem-functional sense it is similar to the surface zones of the oceans, where the plants of the plankton support all the other life forms alongside them and in the depths.

Of course this is an idealized picture of the canopy. Rather than being a uniform layer, based on a uniform age, soil profile, and flat groundsurface, the canopy is usually more like an eroded geological surface where succeeding ones are revealed because the upper strata are not continuously present. Thus the widely scattered emergents are succeeded by a primary canopy layer that is by no means perfectly continuous. The "holes" in this layer are light gaps for lower depths in the canopy that may or may not be defined as separate layers. Knowing the history of a region may provide an explanation for such irregularities. Dimples in the canopy can represent previous treefall gaps or areas of below-average soil fertility.

At the level of the forest floor there are usually two types of vegetation, a woody shrub layer and a herbaceous ground layer, neither of which is usually continuous. The woody shrubs are essentially of two kinds. There are seedlings of forest tree species and specialist ground-story shrubs. The seedlings of the forest species use stored food from their parent fruits for initial growth and then grow very slowly, if at all. Light penetrating to their level of the forest is of very low intensity, and all kinds of physiological "tricks"

Alain Compost/Bruce Coleman Limited

have evolved to capture it. Despite such devices, vigorous growth cannot take place and these are "trees-in-waiting". Their sole chance of reaching reproductive size relies on a treefall in their vicinity. If a light gap opens up nearby they are like runners on their blocks, and they can sprint upwards with a partly formed root system as a headstart device. The other shrubs live very slow lives but can eventually flower and fruit, while the very scattered herbs of the ground layer also live slow sparse lives, exploiting whatever local riches of intermittent sun dapples and nutrients may be present. It is here that parasitic plants may exploit the nutritive resources of the roots of canopy trees to propagate themselves. The giant flowers of the genus *Rafflesia* are external and conspicuous markers of the success of this form of robbery.

At the other extreme, the tight-packed crowns in a complete canopy are an expression of the vigorous struggle for light, of competition for photosynthetic space. Through the process of evolution, this has produced some complex adaptations, strategies, and tactics. Of these the epiphytes of the tropics are among the forest's most characteristic components. In the Neotropical realm alone there may be more than 15,000 epiphyte species—orchids, bromeliads, ferns, mosses, and others that grow entirely without contact with the ground. They use the woody structures of other plants to achieve their position in the sun. Mistletoes, the Golden Bough of European mythology, are common and highly successful epiphytes in many tropical forests. Their seeds are dispersed by birds onto high tree branches and germinate there. They are partial parasites, penetrating the host plant for resources but retaining some photosynthetic powers. In addition to the epiphytes are the epiphylls, which are mainly mosses and grow on the leaves of plants.

Habitats like this riverine forest in Borneo are biologically among the richest in the world. One meticulous survey, of a 50-hectare (125-acre) tract of forest, plotted over 340,000 individual plants with a stem diameter of 10 millimeters (about ½ inch) or more, belonging to about 800 species. The diversity of animals in such a plot is vastly higher.

THE WATER CYCLE

AILA KETO

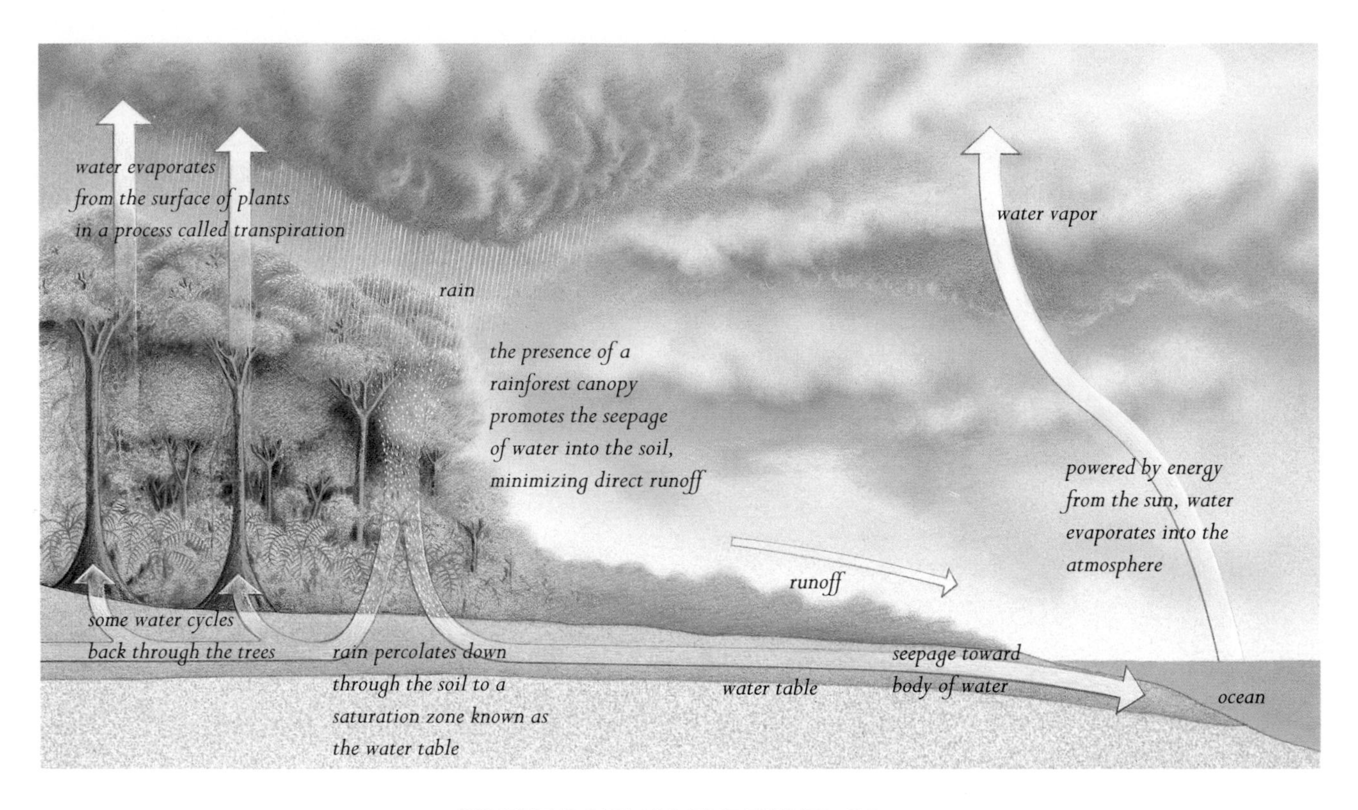

WHERE DOES ALL THE WATER GO?

By far the major part of the Earth's water resources is held within the oceans, the ice caps and glaciers of the Arctic and Antarctic, and deep below the land surface. Less than one-thousandth of the Earth's water is available to support terrestrial life, and this water "cycles" between the Earth's surface and the atmosphere—evaporating from the oceans, lakes and rivers, and from leaf surfaces of vegetation, before returning to the land and the sea as rainfall and snow.

This is the hydrological or water cycle, and it displays its greatest activity where jungles and rainforests grow. In these regions, rainfall is high—ranging from 2,000 to more than 11,000 millimeters (80 to 430 inches) per year—and the dense jungles release water back to the atmosphere at a correspondingly high rate. Rainfall is broken by the forest canopy, and instead of a deluge striking the ground, rainwater drips from the leaves in a fine shower or runs down tree trunks and branches. What happens then depends to an extent on the intensity of the rainfall. In the case of a tropical downpour some of the water runs over the ground and flows into gullies, creeks, and rivers. Much of it, however, seeps into the soil and trickles down through the gaps between the soil particles. Some is held there as soil moisture, while the rest percolates through to enter the subsurface groundwater. The level of this groundwater—the water table—rises and there is a slow and steady lateral flow that eventually seeps out into rivers and oceans. Here, evaporation takes place, and the water returns to the atmosphere to begin the whole cycle again.

The water held as soil moisture is taken up by the roots of the forest's trees and shrubs, and from there it travels upwards through the stem and branches to the leaves. At the leaf surface, water evaporates throughout daylight hours. There is an almost constant flow of water from the soil through the tree or shrub and into the atmosphere. This is known as evapotranspiration, and it is inextricably linked to the process of photosynthesis, in which carbon dioxide from the atmosphere is taken up by the plant and converted to various organic forms including wood. Keeping the surface of the leaf moist allows carbon dioxide to dissolve in the water within the leaf tissues. Both evapotranspiration and photosynthesis are driven by the energy of the sun.

All plants undergo these processes, of course. But, the trees of tropical forests generally excel in them. The large volume of water entering the atmosphere over tropical rainforests as a result of evapotranspiration has a significant effect on regional rainfall. For example, it has been estimated that more than half of the rain that falls on the great jungles of the Amazon River basin returns to the atmosphere through the foliage of the trees, to fall again as rain upon the Amazonian forests. ●

As well, there are lianas and vines that grow upwards, often from light gaps, using the forest trees as scaffolding. This is an extremely successful strategy, and 1 hectare (2½ acres) of forest may contain 1,500 or more lianas.

Some of the most interesting rainforest plants start growth high in the canopy where premium light triggers their growth downward toward the soil. The misleadingly named strangler fig *Ficus crassiuscula* is one such vine. Once it reaches the ground level and takes root it puts on a prodigious growth of woody strengthening tissue. The host tree from which it started its downward growth then dies, shaded out—not strangled—by the fig's vigor. At this stage the fig looks like a gnarled and contorted tree, giving little indication of its extraterrestrial origins.

THE PAST AND THE FUTURE

Before this century the tropical forests of the world covered a much larger area than they do today. From fossil evidence we know that forests grew in what is now Malaysia and Indonesia in Tertiary times (65 to 2 million years ago). They seem to have had a similar flora to today's forests.

Fossils and biogeography can tell us something of the post-Pleistocene (after glaciation; 10,000 years ago) history of rainforests. However, the very inaccessibility that makes present-day forest biology such a difficult subject also militates against the study of paleontology in the same regions. Studies of pollen remains and phytoliths (the mineralized inclusions in leaves, stems, and fruits that show up after the soft parts have decayed) are beginning to yield some results, but for most areas of rainforest there is still nothing but inspired speculation.

It seems that the rainforests of the great realms have shrunk and expanded through the post-glacial period, but have sustained large areas of forest for long periods. Speciation in the Amazon area suggests that at times there were numerous islands of forest in the midst of a sea of grassland. When they again became interconnected, species spread out and the forests finished up with their present richness.

If the great forests are again reduced to a few tiny islands in a sea of human activity, will they ever again be interconnected in a gloriously diverse treasure-house of biology? Probably not, but then the Thames Valley, in Britain, was once the site of tropical flora and fauna that must have approximated a rainforest. It disappeared, but tropical forests survived and proliferated. The big difference between then and now is the presence and activity of humankind. ■

Michael & Patricia Fogden

A three-toed sloth Bradypus infuscatus *in a Panamanian rainforest. Long-lived, slow moving, and solitary, sloths live in trees and feed exclusively on leaves.*

John Cancalosi/Bruce Coleman Limited

Among the largest of all eagles, the majestic and gravely endangered harpy eagle Harpia harpyja *of South America feeds largely on monkeys, which it often captures in extraordinarily agile, high-speed pursuits through the rainforest canopy.*

2 In the Treetops

ANDREW MITCHELL

Ever since naturalists began to explore tropical jungles they have wanted to get into the rainforest roof, correctly believing that a new "continent of life" existed there waiting to be discovered. But the lowland tropical forests of the world are dominated by giant trees 10 stories tall, most without branches for the first 20 meters (65 feet) and whose trunks are often coated with embracing lianas or clinging epiphytes such as ferns and bromeliads that harbor poisonous snakes or stinging insects.

Early Days

For 150 years the difficulties of climbing these trees served as a deterrent to aspiring arboreal botanists and zoologists, although they have conjured numerous ingenious techniques, ranging from rope-firing cannons to specially trained monkeys, shotguns fired at branches, and even the felling of whole trees, as a means of gaining access to the canopy's secrets. The ropes often ended up as so much spaghetti dangling from the trees; the monkeys either ate the fruit samples they were supposed to deliver or simply refused to come down. Shotguns deposited shattered samples onto the heads of upwardly gazing botanists, and felled trees produced merely a crop of immobile plants and insects and the occasional dazed mammal. And nor did using indigenous peoples work. Although they are usually adept at climbing tall trees, they are not always available or willing to work for long periods in such dangerous conditions. As a consequence, the creatures and plants that inhabit the treetops and their influence on the forest ecosystem have remained, until recently, shrouded in mystery.

There are good reasons to explore the canopy. While the interior of the forest is a gloomy place, starved of energy-giving light, its canopy is the real powerhouse. Here, leaves track the sun like millions of solar panels, forging sugars through photosynthesis, mixing nutrients, and maximizing growth. Accordingly, it is here too that the greatest biomass of animals and plants is to be found, although questions about their number, variety, and interconnectedness are still largely unanswered. How varied is life on Earth, and why does it seem to be so diverse in rainforests? Many of the animals that inhabit the canopy, such as flying frogs, gliding squirrels or snakes, tree possums, monkeys, arboreal anteaters, high-flying butterflies, and giant bumblebees journey to the ground so rarely, if ever, that little is known of their lives. Are the bats that are common at ground level also common in the canopy? Do trees act as tenement blocks with a permanent collection of occupiers, or do they merely house itinerant guests?

Luiz Claudio Marigo/Bruce Coleman Limited

A typical rainforest canopy in the Serra Dos Orgaos National Park, Brazil, with Cassia *and* Tibouchina *trees in glorious bloom. Rainforests encompass two very different worlds—one of quiet and calm on the forest floor, where light, temperature, and humidity are virtually constant; and the other high in the canopy above, where light levels range from full sunlight to cathedral dimness, with corresponding erratic extremes in wind, temperature and humidity.*

Trees conduct their sex lives in the privacy of the canopy, their crowns of flowers pollinated by suites of pollen messengers such as bees, birds, and bats. Their fruits are eaten and their seeds dispersed by legions of pigeons, fruit bats, toucans, mice, and monkeys. But who serves whom is not known, and which insect is essential to which tree's flowers is still waiting to be discovered. If we do not know even these simple data, how can we begin to understand how the rainforest reproduces itself? It is at the canopy level that the rainforest also exerts its greatest claim to be a contributor to our atmosphere, drinking in carbon dioxide through billions of tiny leaf-borne pores, breathing out oxygen that the rest of the world may inhale.

But just how important are rainforests in controlling local, regional, or even global climate? Can they help lessen the greenhouse effect by soaking up carbon dioxide? What will happen if they are cleared? Such questions are as compelling as they are fundamental, and to answer them we must explore the canopy of the world's tropical forests.

Michael & Patricia Fogden

Few birds are more characteristic of the American tropics than fruit-eating toucans, such as this chestnut-mandibled toucan Ramphastos swainsonii *peering from its nest in Costa Rica. The function of the toucan's enormous bill is still uncertain, but it is known to be very light and the thin outer shell is supported internally by a spongy web of bony struts and tissue.*

Stairways to Heaven

The first opportunity I had to get into the rainforest canopy was in Panama in 1978 as the scientific coordinator of an international expedition named Operation Drake. This was an appropriate name, as four centuries earlier Sir Francis Drake himself had scrambled up some steps cut into the side of a huge tree in the Serrania del Darien, Panama's mountain divide, and from its crown first gazed out on the Pacific and dreamed of a great enterprise, his circumnavigation of the world. In the early 1970s, while studying the gibbons and monkeys of Mount Mulu in Sarawak, I too had dreamed of exploring an unknown world. In my case that world was the rainforest canopy.

The primates I wished to study remained tantalizingly out of reach, mere silhouettes in a sea of dark outlines against the sky. The pain in my neck from staring upwards into the trees forced me to think of new techniques of studying such creatures, where I could meet them face to face, on their own ground, so to speak—in the treetops. With Dr. Stephen Sutton, an ecologist at Leeds University, I came up with a new design of lightweight, portable aerial walkways that could be strung between the tree crowns like suspension bridges. Starting from tree platforms I imagined teams of scientists walking through the canopy sampling new life forms at arm's length, spending the night there, placing traps for insects, birds, or bats 40 meters (130 feet) above ground, getting some idea of the three-dimensional nature of the forest. The dream came true in Panama when I stood in front of a massive *Couratari* tree up which a rather flimsy rope ladder had been placed by members of the Royal Engineers. They helped build our walkways in Panama, and later in New Guinea, Indonesia, Sarawak, and Costa Rica.

The ladder rose like a stairway to heaven. I put my foot on the first rung. It broke. Undaunted, I began to climb, shakily. I had no particular head for heights. Three-quarters of the way up I noticed I had no safety line and the tree began to move with ominous creaks. This was an unexpected sensation. A light breeze had caught its crown, and with further creaking it swayed back the other way. I clung to the ladder, frozen into immobility. Salty sweat trickled from my eyebrows into my eyes. After this inauspicious beginning I realized that these huge trees are designed to sway, just as skyscrapers are, and completed my climb. Even so, I wondered if monkeys ever had a fear of falling.

My introduction to the canopy was spectacular. Having scrambled into the crotch of the tree 30 meters (100 feet) above ground, I could gaze out over an undulating sea of many colors. Crowns in flower rose like great upwellings of bright yellow, lilac, or red, set in hills and meadows of green leaves rustled by gentle humid breezes. The harshness of the canopy sunlight was a great contrast to the low light conditions of the forest floor. Macaws dressed in scarlet and blue screeched overhead and banked through the green valleys, small vireos and tanagers flitted back and forth between the crowns. An emerald-green hummingbird inspected my red T-shirt not 60 centimeters (2 feet) away and then whirred across the canopy probing red vine flowers for nectar.

The branches of the tree in which I was sitting were festooned with epiphytes, botanical hitchhikers seeking a place in the sun. Ferns, bromeliads, mosses, orchids, lichens, and liverworts crowded the branches, each a miniature jungle and home to numerous creatures that scientists had perhaps never set eyes on before. Even the *Couratari* tree in which I sat turned out to be a new species. Such is the unexplored nature of these magnificent environments that to discover an undescribed species of so vast a size is not unusual.

Since that time, two decades ago, a new breed of arboreal naturalist has been born, and individual botanists and zoologists in Central America, the Amazon, Africa, Asia, and Australia are beginning to creep off the forest floor and enter the rainforest roof. It is an interesting time. After all it is 15 million years since our ape-like ancestors left the canopy to evolve into a new life form on the open plains. Now their descendants, furnished with a little more intelligence but perhaps no better a vision, are climbing back there. What have we found?

Opposite. *The orang-utan ("wild person" or "man of the forest" in Malay and Indonesian) is restricted to the forests of Sumatra and Borneo, where it is seriously endangered because of rainforest logging. Fully adult males may grow to 90 kilograms (about 200 pounds) in weight, but they are very agile and spend almost all of their time in trees. Unlike other apes, the orang-utan is an essentially solitary creature.*

Highways in the Trees

First, the canopy is not a haphazard mishmash of branches—a jungle of mere human imagination. It is in fact a network of arboreal highways as familiar to the animals living there as any favorite walk is to us. Highways are instantly visible in the canopy because of the number of feet that use them. Squirrels, monkeys, or mice pat down the moss that coats branches, forming an avenue through the larger plants, which grow up on either side.

That these trackways are memorized is shown by the fact that a squirrel being pursued will rush headlong into space from the end of a broken branch it knew to be there the last time it passed. At home even in such a situation, the squirrel will merely stretch out a paw as it plummets earthwards and clutch a twig along which it will then scamper as if nothing had happened. Monkey troops will regularly use the same pathways linking trees on their way to crowns in fruit. In the American tropics many monkeys such as howlers and spider monkeys have prehensile tails, which enable them to hang from branches and gain access to choice leaves, or to fruit that is denied to other species. In Africa the pangolin and silky anteater have similarly prehensile tails.

Also, contrary to expectations, tree crowns do not interlock but, in a sense, rub shoulders across a gap of about 1 meter (3 feet). The reason for this might be supposed "aggression" between trees of different species abrading each other's crowns in storms, or the need to avoid infection from caterpillars exploiting a neighbor's crown. But whatever the explanation, the animals of the forest must overcome these gaps through the use of powerful back legs, gliding membranes, or wings. The saki of Colombia are not known as "monos voladores", the flying monkey, for nothing. These monkeys are capable of leaping 10 meters (more than 30 feet) from one tree to another, which is a big advantage in more open forests.

Beneath the canopy itself, in the understory, live the clingers and leapers such as tarsiers, marmosets, and lemurs, which jump sideways from trunk to trunk. Here, gliding squirrels perform best, leaping on cloaks of skin spread between outstretched limbs. The flying lemurs of the Philippines and Malaysia are capable of gliding 100 meters (more than 300 feet) across a valley to a favored feeding tree. The flying frogs of Southeast Asia are less skillful but can make an escape from a high leaf, glide on four parachutes of skin stretched between the toes of each foot, and can even move with some direction by altering the configuration of their feet in flight. The Malaysian gliding snake *Chrysopelea pelias* does not so much glide as plummet, but

Jean-Paul Ferrero/AUSCAPE

flattening its body sideways by pushing out its ribs and lateral sinuous movements combine to give direction to its fall, which through evolution has made the difference between life or death in escape. However, the master reptilian aviator is the flying dragon *Draco volans*. An inconsequential lizard about 20 centimeters (8 inches) long, it launches itself from high branches to glide impressive distances on wings of skin stiffened by its ribs, which can be swung sideways into position or held neatly by the side of its body when at rest.

Flight Tunnels

Birds of course have no problem negotiating tree crowns, but their flight patterns through the branches are also carefully memorized. Flight tunnels exist through the canopy, which speed their journeys, and many canopy hunters such as owls and small hawks have short stubby wings that give them great power and agility in tight spaces. Most mammals cannot fly, but bats can and they also use these flight tunnels when feeding at night.

Small insectivorous species of bats search for their prey among the branches in total darkness by using their high-frequency echolocation sound systems. They emit a pulsed beam of sound through their nostrils and listen as the tone of returning echoes reveals the direction and speed of, say, a moth in front of them. Due to a Doppler effect, the sound waves are compressed if the moth is approaching or lengthened if it is moving away. Some Doppler bats prefer to hang from a branch in the canopy with plenty of space around them and to listen out for insects there. They occasionally launch out into the darkness and snatch insects from the air, some species by scooping them into their mouths with their wings or tail flaps, others by catching them direct in their mouths.

Fruit bats such as flying foxes from the Old World forests of Africa, Asia, and the Pacific prefer to use sight to locate food and tend to set off from their canopy roosts just before darkness falls. The largest species have wingspans of more than 1 meter (3 feet), will travel 20 to 30 kilometers (10 to 20 miles) to find tree crowns in fruit, and gorge themselves noisily through the night before returning in long lines across the forest to their daytime roosts. One of the largest flying foxes, the Samoan fruit bat *Pteropus samoensis*, differs from most other species in the same genus by living alone and flying during the day. I have even seen them soaring on thermals like huge birds of prey high above the forest. Sadly, this species is now highly endangered, as are many other Pacific flying foxes, because hunters shoot them, freeze them, and send them to Guam in the Mariana Islands, where they are considered a delicacy.

The leaves of a tank bromeliad form a watertight bowl that accumulates a pool of rainwater high in the treetops. Related to the pineapple, tank bromeliads are epiphytes but not parasites, and they obtain their nutrients from the miscellaneous animal detritus that filters down to the bottom of the pool. The water-filled bromeliad is also home to a number of small creatures; and is visited by many more.

An alligator lizard pauses to drink at this convenient arboreal filling station.

THE MINIATURE WORLD OF A TANK BROMELIAD

Velvet worms have been discovered in bromeliads, far above their usual home on the forest floor.

The bromeliad pool is a perfect breeding ground for mosquitos.

A female poison arrow frog reverses into the water-filled bromeliad to deposit the tadpoles she has carried up from the forest floor. The tadpoles feed on algae and mosquito larvae in the tank, but to ensure that they do not run out of food the tiny mother frog returns each day and deposits a single unfertilized egg into the bromeliad.

A mouse opossum plucks a damsel fly larva from the bromeliad pool.

Not always safe in their watery nursery, tadpoles can fall prey to developing water beetle larvae.

A variety of large and colorful land snails forage and hide between the leaves.

Predatory and scavenging ants patrol the area for anything they can find.

INSECTS OF THE CANOPY

TERRY L. ERWIN

Species richness, or biodiversity, seems almost limitless when one ponders samples of beetles and other insects collected from the treetops of the Amazonian rainforest. One canopy in Peru with 500 cubic meters (17,500 cubic feet) of foliage (about equal to a two-car garage) will have more than 50 species of ants, 1,000 species of beetles, as many as 1,700 arthropod species, and a total fauna of 100,000 or more individuals.

Samples of beetles from treetops in Panama, Brazil, Peru, and Bolivia (each with more than 1,500 species to compare) reveal that only 3 percent of the species are shared between the four sites. Among three sites in Peru, a mere 310 kilometers (500 miles) apart, the comparison is the same: 3 percent shared! How many species are there? Only 1.5 million life forms are described in scientific literature, while several hundred new descriptions (mostly of insect species) are added annually by a small group of taxonomists using specimens found in museum collections. The canopy species are known from only a few hundred collections, and to date, these few samples are little studied. Nevertheless, some estimates of species numbers range from 30 to 50 million, other more conservative estimates range from 5 to 10 million.

Climate in, and slightly above, the canopy is extremely important in regulating who can live where. But the undersides of leaves do help shield many canopy insects from the sun and rain. Also, because the crowns of tropical trees are very good dispellers of the wind's force (each branch sways in a different direction, vectoring the forces on the main trunk to near neutral), insects need only be sway-adapted. The undersides of leaves also provide insects with relative security from predators. As well, crypsis and mimicry—the ruse of blending into the surroundings by having similar colors and patterns, or by looking like something else—are more the rule in the canopy than the exception. Paradoxically, so too are displays of gaudy colors and patterns that advertise the presence of toxic, or at least bad-tasting, chemicals. Other less- or non-toxic species have evolved similar colors to gain protection from predators with memories of bad-tasting colorful models.

Most of the individuals and biomass in the canopy are ants; most of the species are beetles; book lice and flies are abundant and are represented by many species. The average size of a beetle in the Amazonian rainforest canopy is about 3 millimeters (1/8 inch), and the largest arthropod species are herbivores, caterpillars and katydids, all of which eat the most abundant canopy resource, leaves. Many trees develop chemicals that are stored in the leaves or seeds and protect them from being eaten by insects. But many types of insects, certainly among the most adaptable animals on Earth, evolve to handle the poisonous diet; then the trees evolve to form new chemicals to keep ahead of the insects: this is what Paul Ehrlich and Peter Raven have described as the "evolutionary arms race".

Scientists of the next century face not only the task of documenting the richness of life on Earth, most of which is in the treetops, but also countless discoveries of how biodiversity is assembled into discrete, or often interconnected, ecosystems that sustain themselves for eons if they are left alone. •

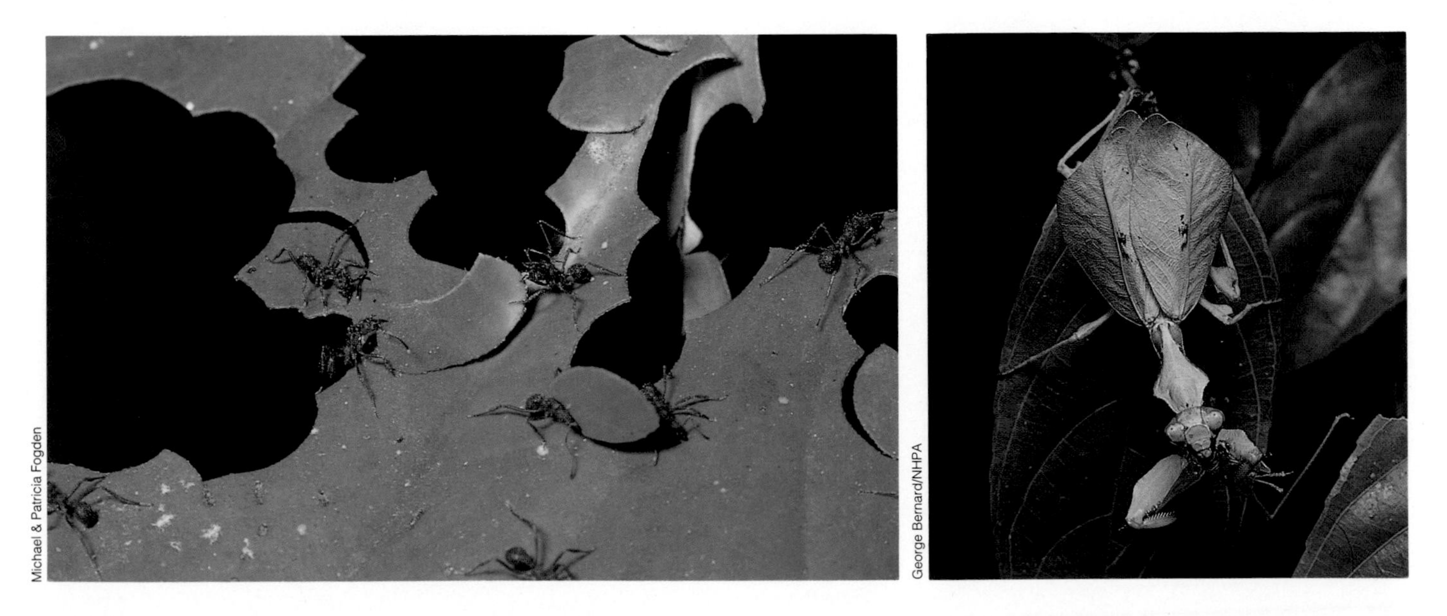

Leafcutter ants collect leaves from the canopy and bring them back to their ground nests, where they are cultivated into a fungus, the ants' staple diet.

Common in the canopy, mantids are predatory insects that occur in a range of sizes, colors, and shapes. This frock-coated mantis is dining on a katydid.

K.G. Preston-Mafham/Premaphotos Wildlife

A spider Cupiennius coccineus *(family Ctenidae) from Central America with its prey, a harlequin-patterned treefrog* Hyla ebraccata. *Rainforest warmth and humidity foster extremes in size in many arthropod groups, and some tropical spiders are large enough to be formidable predators even to small vertebrates such as frogs.*

Eating and Being Eaten

The canopy represents an enormous banqueting table, which its inhabitants exploit to maximum advantage. This is reflected in their territories and in the complexities of their behavior. For example, howler monkeys eat leaves, and because leaves are everywhere they have no need to protect their resources and do not establish fixed territories. But they howl in the early morning so that each group can establish its whereabouts in the canopy, thereby avoiding confrontations.

Sloths, both the two-toed *Choloepus* species and the three-toed *Bradypus* species, have perfected an all-but-sedentary lifestyle and move with glacial slowness, feeding on 20 or so favored species of trees while hanging upside down from their branches. Many tropical trees have evolved leaves that are filled with toxins to ward off insects and other leaf predators, or that are extremely tough and almost indigestible for the same reason. Sloths and other herbivorous primates have, in turn, overcome this by employing bacteria in their guts. But a compost heap in your belly produces little energy, and sloths must therefore conserve theirs. So the sloth is slow and spends anything up to 18 hours a day hanging beneath branches, apparently asleep.

Death for a sloth comes on the silent wings of enormous harpy eagles. The eagles look out for them from the tops of emergent trees as sloths climb up into crowns to snooze in the morning sun. The harpy eagle, on spotting one, dives beneath the canopy to avoid being seen, and swoops up underneath the sloth, at the last minute turning over and snatching it from its

Jean-Paul Ferrero/AUSCAPE

At rest on a tree trunk or limb the flying dragon Draco volans *is well camouflaged, blending in with the browns and grays of the bark. When in "flight" it unfolds the membrane between its protruding ribs and reveals a striking pattern of vivid red.*

branch on talons as broad as a man's hand. Sloths, being unable to run, have evolved camouflage to avoid this end. Their long white hairs are grooved and filled with green algae, which enables them to blend with their surroundings. The ruse works. Sloths are among the most numerous mammals in Central American rainforest canopies, but they are rarely seen.

Whoops, Screams, and Whistles

Food resources that are not available all the time, such as fruiting tree crowns, are often defended by the animals that use them. The gibbons of Asia differ from most leaf-eating primates by living in small family groups rather than troops. They are also the most tuneful of primates, and for a combination of reasons. Each morning the parents sing, the males usually beginning just before dawn. Later, they are joined in a duet by the females. Their magnificent whoops and screams, unique to each species, serve both to enhance the pair bond that unites the male and female core of their family and to delineate their territory.

Each group remains in a large but fixed territory giving them access to the fruiting trees on which they depend, excluding other gibbons. Since they must travel far to reach trees that fruit at different times in their territory, they have also evolved the ability to move at speed through the forest by swinging rapidly beneath branches. They propel themselves across 10-meter (30-foot) gaps with ease, occasionally bounding off boughs and trunks with breathtaking agility.

All the great rainforests of the world have monkeys with loud voices. The howlers of the Old World, the gray-cheeked mangabey of West Africa, and the gibbons of Asia all call in the early morning—before it gets too hot to bounce their cries between the cooler air trapped within the forest and the warming air above, and to avoid the background noise of the day's bird and insect song that might otherwise drown their voices. Only those primates with low-frequency calls, such as orang-utans and siamang, call later in the morning when the volume of their call and its low frequency punches it through the forest. It is interesting that the long-distance calls of most rainforest primates are about 200 hertz, a sound window that penetrates the forest best.

Rainforest birds employ similar techniques. Many living in the forest understory call with low-frequency booms, while others use pure tones and whistles whose simple messages cut through the hum of insects and are less easily soaked up by sound-absorbing vegetation. As you climb into the canopy it is noticeable that the birds there use much more complicated songs, employing shrill notes that convey greater amounts of information.

Martin Dohrn/Bruce Coleman Limited

Gifts for Pollinators

Trees also need to pass their genes around and disperse their offspring to germination sites, but because they cannot move they have evolved to seduce help from the animal world and the elements. In temperate forests strong winds are often all that is needed, but in the rainforest little wind penetrates the understory and gusts and squalls are relatively infrequent over the canopy. Those few rainforest trees that rely on the wind to disperse pollen or seeds are either pioneer species that spring up in light gaps after a treefall or grow at the edges of the forest where wind is most likely to affect them. Most trees in the tropics are dioecious, meaning that they require their pollen to be carried, often over quite long distances, to other trees of their kind. In the rainforest canopy they "seduce" their pollination messengers through color and with gifts of nectar and pollen.

Widespread in the Old World tropics, fruit bats maintain an intricate relationship with the flowering and fruiting cycles of many rainforest trees. Most fruit bats forage by night and roost by day in trees, but members of the genus Rousettus, *like this Indonesian species, differ strikingly from other fruit bats in that they habitually roost in caves.*

In the understory few flowers are apparent. Often they are red, such as *Brownea rosa-de-Monte*, or orange and yellow, such as *Heliconia's* spiky, cockatoo-crest bract. The bracts of "hot lips" flowers look like a human pout. Most of these are brightly colored to attract hummingbirds or trap-lining bees, which can locate them easily against the background green of the forest. However, identifying just who the visitors are to the flowers of most tall rainforest trees is one of tropical biology's most exciting challenges. While walking through the forest the only indication that a crown high above your head is in flower is a carpet of yellow or lilac flowers scattered across the forest floor, their stamens lying useless, their pollen gone. But who might the pollinators have been?

The flowering cycles of tropical trees are themselves something of a mystery. In lowland evergreen forests most seem to have no obvious pattern. Seasonal changes, such as a dry spell, usually seem to trigger flowering, but several years may be missed. One tree in Southeast Asia seems to flower only after being bathed in ozone from lightning strikes. *Tachigalia* trees in South America have some kind of internal clock enabling them to flower all at once and then die en masse once they have fruited. In Malaysia and Indonesia an astonishing phenomenon occurs every five to six years, when the whole canopy comes into bloom. The tree species which dominate these forests, known as dipterocarps for their two-winged seeds, flower in a staggered sequence, producing richly perfumed tiny white flowers at night that until recently were believed to be pollinated by the wind. In the 1970s, however, scientists led by Peter Ashton from the University of Aberdeen suspended themselves from a bosun's chair attached to a long boom and discovered the flowers to be full of tiny thrips, insects barely 2 millimeters (1/16 inch) long. Pollen grains coated with oil stuck to the thrips and were then dispersed on the wind when the weakly flying insects left the flowers during the night.

Smallest and most agile of the apes, gibbons live in Southeast Asia. Their preferred style of locomotion is brachiation—that is, dangling by their arms and swinging hand over hand through the trees.

To be sure that their pollen is accurately delivered, it is to the trees' advantage to have a personal pollen messenger such as a large insect, bird, or bat that is powerful enough to find another tree species of the same kind. Bees such as those in the genus *Eulaema*, almost 5 centimeters (2 inches) long, make ideal messengers for many canopy trees, whose flowers often appear designed to "fit" them by depositing pollen at a particular position on the bee. Such bees may be carrying the pollen of four or five different tree species as they speed over the canopy, all positioned at specific points on their bodies, head, thorax, or abdomen. The design of the receiving flowers ensures that they pick up only the pollen from their own species.

This makes for far more efficient pollination than reliance upon the vagaries of the wind, and canopy flowers often suit their customers' size and behavior. Nocturnal flowers may be white, making them more visible. Those visited by bats are large and tough enough to provide a landing stage, and they produce their nectar and pollen in copious amounts to suit the bat's appetite, whereas flowers visited by bees produce smaller amounts of highly concentrated nectar. Flowers visited by hawkmoths attract them from miles away because of their sickly scent. They open only at night, and the moths probe them with a long proboscis, as the flowers have no landing stage. The proboscis of Madagascar's hawkmoth *Xanthopan morgani* is 25 centimeters (10 inches) long, and no other hawkmoth can drink the nectar of the orchid it exclusively pollinates.

Chains of Dependence

Such complex interactions between animals and plants are a feature of tropical forests and are one of the factors that make them such fascinating places to study. The animal that pollinates a tree may not be the one that disperses its seeds. One tree may depend upon a whole series of creatures to complete its life cycle—a bee to pollinate it, a bat to disperse its fruit, a rodent to bury the seed on the ground where it may have fallen. Conversely, the crucial bee, bat, or rodent may in turn depend upon a whole series of different

trees throughout the year for its food. Disrupting just one part of any of these chains of dependence can spell doom for the whole system.

There are of course some checks and balances built in, but it is because of the immense complexity of these kinds of interactions in the tropical forest ecosystem that no one yet understands how to regrow a rainforest. Much of the information that we need remains to be discovered in the canopy.

Tall trees act as a matrix upon and within which the rest of the forest lives, and their canopy is remarkable for the vast but as yet uncharted numbers of animals and plants that depend upon them for support. Epiphytes such as bromeliads, ferns, and mosses have neither trunks of their own nor roots in the soil, so they grow direct from branches high above ground. Their seeds are usually light enough to float from tree to tree on the wind. Other seeds are deposited high in the trees, often by birds. Mistletoe berries have a laxative in them, and the seeds they contain are coated with a sticky substance; they pass rapidly through the flowerpeckers that eat them although the seeds stick to the tail feathers and the little birds then skitter up and down rubbing their bottoms on branches to be rid of them. In the process the seeds are deposited exactly where they need to be.

Left. *Brachiation, such as commonly used by the orang-utan, is an effective means of locomotion in the trees, but puts a premium on long arms, curved hands, and exceptionally mobile shoulder joints—characteristics that might not be so useful in certain other situations, such as walking upright or the dextrous manipulation of small objects.*

Miniature Swamplands

To cope with lack of water many epiphytes have evolved tough water-conserving skins similar to those of desert plants. The bromeliads of the New World tropics, relatives of the pineapples, may retain up to 10 liters (18 pints) or more of water in a central "bucket" of overlapping leaves. This airborne water supply is very valuable to the animals that live in the canopy. Monkeys, lizards, and mice will pause at these arboreal filling stations to drink. Mosquitos and giant damselflies will use them as breeding grounds; even earthworms have been discovered working their way through this miniature swampland, far above the earth in which we usually expect to find them.

Perhaps most extraordinary are the small ground-living frogs that use bromeliads as nursery grounds for their tadpoles. These poison arrow frogs of the genus *Dendrobates* lay a small number of large eggs in burrows on the forest floor, which are then guarded by the brightly colored adults. When the tadpoles hatch they are encouraged to slither onto the female's back, and she begins a momentous journey. First she reaches a tree trunk and then climbs through the moss and plant growth on the bark until she locates a branch. She searches for a bromeliad filled with water and reverses into it, depositing each tadpole into a different bract and some into different bromeliads. Here the tadpoles will feed on algae and mosquito larvae, but to ensure that they do not run out of food the tiny frog returns each day and deposits a single unfertilized egg into the bromeliad.

With the aid of its prehensile tail, a sort of fifth limb, the spider monkey can walk along a branch or dangle below it with almost equal facility.

Recycling Machines

It has recently been discovered that epiphytes play an extraordinarily important part in the nutrient recycling of the forest. Rainforests are the ultimate recycling machines. The soil on which they stand is often shallow and infertile, but the rainforest sips energy from the sun and almost feeds upon itself, taking in minute amounts of nutrients from airborne dust and minerals from the soil. Epiphytes help to trap this dust and capture leaves from the air, creating a mulch of decaying leaves that sustains them in the absence of soil. This is so valuable a food resource that trees have been discovered to grow roots out of their branches to tap this nutrient lode, whose value is further increased by the numerous ants that colonize hollow branches and epiphytic ant-plants.

One tree may have 20 or so *Myrmecodiea* ant-plants providing hanging apartments for an amorphous colony of tiny *Iridomyrmex* ants composed of many queens and countless workers. The workers gather decaying insect carcasses from the forest floor and store them, along with their own dead, in minute canopy graveyards inside the plants. The decaying matter produces carbon dioxide and nutrients for the plant, while the ants get a home and probably grow fungi on the debris which they then eat.

This kind of nutrient recycling on which the rainforest depends—above ground, from leaves to soil, to fungal decay, to roots and back into leaves—is known as a closed system. It is quite unlike the open nutrient-recycling process of many temperate ecosystems in which substantial inputs from deep soils are available. This is one of the reasons why rainforests are so vulnerable and why they are usually so unproductive as farmland once the forest has been removed. Threats from global warming and the greenhouse gases are now causing scientists to examine the world as one giant closed system that may itself be highly vulnerable to change.

The tiger moth Amaxia flavipuncta *of Amazonia. Named for their mostly black and yellow coloration in warning to predators of their toxic blood, tiger moths of many species occur in forests and woodlands almost around the world.*

Kjell.B. Sandved/Oxford Scientific Films

A granular poison arrow frog Dendrobates granuliferus *climbs a tree with its tadpole on its back. In temperate regions most frogs ignore their eggs, but many tropical rainforest species invest considerable parental care in their young—care that may involve feeding, guarding, or transporting them, depending on the species. This trend reaches a peak in the poison arrow frogs of the family Dendrobatidae, which exhibit the most complex and varied reproductive behavior of all amphibians.*

"Off-the-shelf" Technology

World forests are believed to play an important part in mitigating climate, but the scale on which they do so is largely unknown. They produce oxygen and absorb carbon dioxide. The canopy releases moisture that has fallen into it through transpiration, from roots to leaves to air, but providing reliable data for climate models is almost impossible without accurate measurements from the canopy. An exciting initiative to obtain such accurate data for the first time has just been begun by Dr. Alan Smith, Assistant Director of Terrestrial Research at the Smithsonian Tropical Research Institute on Barro Colorado Island in Panama. The means to do this is a giant crane whose thin tower ensures it can be erected between the trees without damaging the forest.

Entering the canopy by crane was very different from my first climb up on a rope ladder. The huge construction arm merely lowered its basket through the branches to the forest floor, and with a wave to the operator above me I was hoisted above the canopy in

Raphael Gaillarde/Gamma

complete safety and comfort. From here Alan Smith and I were able to drop in on any part of the forest at will, moving out along the revolving 30-meter (100-foot) jib or lowering the cable from which the cage was suspended. The outermost limits of the canopy were at last easily accessible. The first measurements of the photosynthesis and transpiration of the forest at multiple points in the canopy are being made using this "off-the-shelf" technology. This technique and others, such as the massive netting Sky Raft of Professor Francis Hallé, which can be lowered into the treetops and moved about by a purpose-built airship, are opening up nature's last biological frontier to comprehensive scientific investigation for the first time. These technical developments are as significant as researchers' first attempts to plumb the ocean

Until quite recently the rainforest canopy has resisted attempts to explore it. Dangling from rope ladders or similar climbing apparatus is possible but strenuous, inefficient, and dangerous. Here an airship is used to position one of the most innovative and colorful devices used to date, Professor Francis Hallé's Sky Raft. Sky Raft comprises a huge wheel of inflated polyethylene tubes connected by a web of rope netting, the entire structure "floating" on the canopy surface. It enables researchers, supported by rope webbing and anchored by safety harnesses, to move more or less at will over a large area in the canopy, and at the end of the day even serves as a surface upon which to sleep. Rope ladders are the only contact with the ground far below.

depths with deep-sea submarines, and as the diversity of life in the canopy is far richer than anything discovered beneath the sea, scientific findings are likely to be of even greater significance.

Since studies of the canopy began a decade ago the estimate of species inhabiting the Earth, especially of insects, has increased from about one million to 30 million—some even argue for 50 million. This supreme demonstration of our ignorance of the planet and how it works shows how shortsighted humans can be, because we are destroying the biodiversity of the rainforest just as the true significance and value of what is there is being revealed. It is as though the door to a vast and magnificent library is being opened just as the flames are spreading across its shelves. ■

GARDENS IN THE AIR

PETER BERNHARDT

Evolution has designed many plants to pass most of their lives clinging to the branches of trees and stout shrubs. Such plants are properly called epiphytes (from the Greek epi *meaning "upon", and* phyte, *"a plant"). Since young epiphytes must expose their new roots and delicate crowns to the elements, few survive in true deserts or in regions experiencing seasonal freezes, and the vast majority of the 29,000 or more epiphytic plant species are restricted to the wet tropics. They account for 35 to 63 percent of all plant species in some wet forests of Central and South America and may contribute up to 45 percent of the total leaf biomass.*

Epiphytes don't attack their host trees and rob them of water and nutrients. However, not all trees provide suitable perches. For example, comparatively few epiphytes can colonize the smooth, vertical trunks of palms or the densely shady limbs of rubber trees. Trees with cracked or wrinkled bark bearing microgardens of moist lichens and mosses seem to offer superior nurseries for the incoming seeds and spores of epiphytes, which are forced to produce far more offspring than their closest ground-dwelling relatives because so many of their seeds and spores fail to "hit the right targets". Epiphyte seeds released on the wind, many of which are microscopic in size, are equipped with thin glider wings or hooked parachutes. Even those epiphytes that offer fleshy fruits to birds, bats, and primates may pack up to 1,000 seeds inside each small berry. Such seeds tend to be sticky, and the animal is forced to wipe its beak, mouth, feet, or anus against a twig to remove the gluey irritants. This behavior "plants" the next generation.

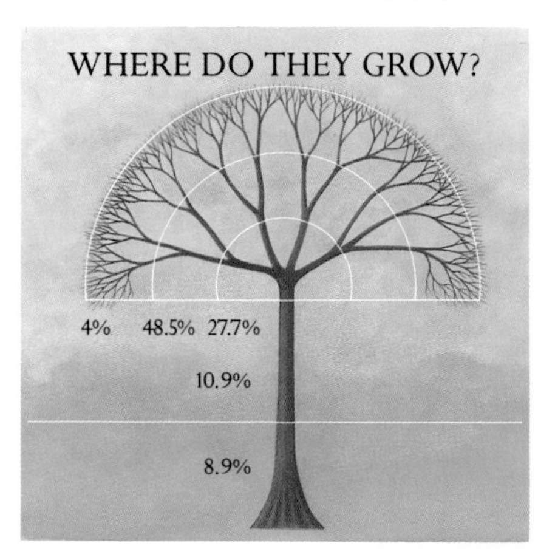

This diagram shows the different zones in a host tree and the percentage of epiphytic orchids found in each of those zones in a West African forest.

Epiphytes have been called air plants but they really don't live on air. They live on what the air brings them: dust, organic debris, and water. Bacteria and fungi break down dust, fibers, dead leaves, and animal waste until they form a protective mulch around an epiphyte's exposed roots and stems. This layer of humus keeps water in and contributes trace amounts of mineral salts. A common growth pattern in some epiphytes is for the roots to grow upwards and outwards like fingers and for leaves to bunch together to form tangled baskets. These epiphytes have become "trash basket" plants, catching the litter of trees and their animal occupants. The richest epiphyte zones tend to be found in middle altitudes of tropical mountain slopes (for example, at 1,000–2,000 meters/3,200–6,400 feet, in the northern Andes). At such altitudes epiphyte roots and leaves receive daily baths of fog and mist from low cloud banks but are not subjected to frosts.

This ant-plant shows the interconnecting cavities that make this type of epiphyte a suitable home for countless numbers of ants; the plant also benefits from the association, being able to absorb nutrients from the debris collected and stored by the ants.

Epiphyte species are found in 83 families, and the majority belong to flowering plants and true ferns. A plant family seems more likely to evolve epiphytes if it has a history of producing fleshy leaves and stems that can store water and a tolerance of gloomy shade and mineral-poor, boggy, or acid soils. Upon viewing the wet forest one is struck by the unique mixture of unrelated plants sharing epiphytic habits of survival. Bromeliads, begonias, peperomias, and the resurrection plants (genus *Selaginella*) are some of the best-known epiphytic groups. Less familiar epiphytes include some species of carnivorous bladderworts (genus *Utricularia*), hare's-foot ferns (family Polypodiaceae), true cacti (for example, genus *Rhipsalis* and genus *Epiphyllum*) as well as many wild tropical relatives of the familiar African violets (family Gesneriaceae), arum lilies (family Araceae), and the blueberries and rhododendrons (family Ericaceae).

Orchids are the most successful epiphytes. About 18,000 orchid species have scientific names but it's estimated that an additional 10,000 still await description. Nearly 70 percent grow only as epiphytes. Every wet tropical forest contains epiphytic orchids, with the greatest variety of species found in the

mountains of South America and Southeast Asia, where dozens of new species are described annually. Why do orchid epiphytes dominate? Orchids produce the smallest seeds of any family of flowering plants. The pod of an epiphytic orchid may release seeds by the hundreds of thousands. Such seeds are measured in microns, and the transparent seed coat enfolds the simple embryo like a balloon. The seeds drift away from the parent plant like tiny blimps, often for long distances. Small wonder that in the nineteenth century some mainland orchids recolonized the Indonesian island of Krakatoa less than 20 years after a volcanic explosion sterilized the island of life. Once they have germinated, orchids seem to be well adapted for life in the treetops, as their roots wear an absorbent sheath known as the velamen. This loose, spongy skin adds extra surface area, which increases the root's ability to absorb rainwater dripping down a branch. Root cells host the coils of fungal threads, and these threads extend beyond the orchid root and digest debris, collecting nutrients for their orchid "landlord".

Epiphytic orchids can withstand limited drying as their secondary stems develop into swollen canteens known as pseudobulbs.

The relationships between the flowers of epiphytic orchids and insects are among the most specialized in nature. For example, about 250 species of angraecoid orchids of Madagascar conceal their nectar at the base of long, hollow spurs. The white flowers release a perfume at night that attracts several species of sphinx moth. The moths drink the nectar and inadvertently carry away wads of orchid pollen on the base of their long tongues. Thousands of orchids throughout the tropics produce drab flowers the size of pinheads smelling of rot or mildew. These little stinkers may offer nectar droplets to tiny drosophilid flies, encouraging cross-pollination, or the flies may have been lured by false promises.

Finally, most of our corsage orchids descend from wild epiphytes of Central and South America. Their colorful flowers secrete spicy fragrances instead of nectar and are pollinated only by male bees in the tribe Euglossini. The male bees carry "perfume flasks" on their hind legs instead of pollen baskets, and they pollinate the orchids in exchange for collecting scents. The male euglossines use the pungent compounds to manufacture a sort of "bee cologne" essential to the completion of their own life cycles.

There are many more untold tales about epiphytes. Fieldwork conducted in the canopies of Costa Rica, Panama, and the Andes is especially intense and is funded by botanical gardens throughout the world. New species are described in botanical journals each month. Research on the life cycle of epiphytes ultimately enriches our greenhouses, contributes to our medicine chests, and, most important, defines the limits of a healthy planet. ●

Above. *Epiphytic orchids, such as this* Dendrobium *from the New Guinea highlands, colonize various parts of their host trees. A few species live on the outer twigs and lower trunk whereas the majority thrive on the crotches of older branches.*

Left. *This small portion of a tree in a Costa Rican rainforest shows the abundance of epiphytes that typically colonize the branches and trunks of tropical rainforest trees. Epiphytes are a substantial component of the total leaf mass in tropical rainforests.*

BIRDS OF THE CANOPY

FRANCIS H.J. CROME

Some of the most beautiful birds on Earth frequent the rainforest canopy. Streamers of brilliant lorikeets, chattering as fast as they fly, appear and disappear like a jewelled express train. Great hornbills honk from some high dead branch. Toucans, macaws, hummingbirds, birds of paradise, or quetzals epitomize tropical bird life.

The habits of canopy birds are as varied as those of birds elsewhere in the forest, and most of them occupy a broad vertical range in the forest profile. Many, such as the brilliant red and black minivets of Southeast Asia and the tyrant flycatchers of South America, will follow the outside of the canopy foraging at the forest edge or in large gaps quite close to the ground. Canopy birds often form mixed-species foraging flocks. In South America these flocks often consist of a dozen or more species of colorful tanagers and several flycatchers; in New Guinea and the Pacific they may be honeyeaters, flycatchers, and warblers, or perhaps pitohuis, cuckoo-shrikes, and birds of paradise; in Southeast Asia the flocks would probably be made up of bulbuls, flycatchers, and babblers.

Cuvier's toucan Ramphastos tucanus *is a member of a family of about 40 species of rainforest birds found only in Central and South America. All are strictly arboreal, nest in tree cavities, and eat mainly fruit. With a body length of about 58 centimeters (23 inches), Cuvier's toucan is one of the largest of its family.*

Stephen Dalton/NHPA

Because more leaves mean more insects, the most common canopy birds are insect eaters. But the canopy also provides nectar, pollen, and fruit, and entire families of birds specialize on these resources. The major groups of nectar feeders include hummingbirds in South America, sunbirds in Africa and Asia, and honeyeaters in the Pacific. These birds are tiny to medium-sized, the smallest hummingbirds (the woodstars) being about 6.5 centimeters (2½ inches) long.

The frugivores (fruit eaters) are, however, perhaps the most characteristic group of canopy birds. In the Neotropics toucans are the major frugivorous species, although they may also eat young birds and eggs. The 12 species of *Ramphastos* toucans are perhaps the largest and the most spectacular of the canopy birds. In Southeast Asia and Africa hornbills replace toucans, and their incredibly loud cries are characteristic of the Indo-Malayan forest, where more than half of the 45 species of hornbill are found. They are huge birds—the yellow-casqued wattled hornbill *Ceratogymna elata* of West Africa can weigh up to 2 kilograms (4½ pounds), and the helmeted hornbill *Rhinoplax vigil* of the Malay Peninsula, Borneo, and Sumatra is 1.5 meters (5 feet) long.

In New Guinea, Australia, and the Pacific the major canopy frugivores are fruit pigeons and birds of paradise. Some birds of paradise are terrestrial or occupy the substages of the forest, but the best known canopy frugivores are the seven species of "true" birds of paradise belonging to the genus *Paradisaea*. They are gregarious birds, and they often join other species in mixed foraging flocks. Except for the blue kumul *P. rudolphi*, which displays solitarily in its own area, the males form "leks" and display communally in favored trees. In these elaborate shows they flaunt their cascades of flank plumes to entice females to mate.

Unlike some pigeons, the smaller canopy-dwelling fruit pigeons are spectacular in color—bright greens, brilliant reds and oranges, yellows, purples and pinks,

Scarlet macaws Ara macao *in flight. Until recently these vivid and spectacular birds were a common sight throughout much of forested Central and South America, but their numbers have now been seriously depleted due to habitat clearance and trafficking for the pet trade.*

blues and golds. The orange dove *Ptilinopus victor* of Fiji is fiery orange, and one of the most beautiful. Unlike toucans, hornbills, and birds of paradise these pigeons eat virtually only fruit.

Parrots are common canopy species in nearly all the world's rainforests, and they range in size from the huge macaws of South America to the tiny hanging parrots of Asia and the equally tiny pygmy parrots of New Guinea. New Guinea perhaps has the world's greatest diversity of parrot species and possibly the brightest and most colorful.

The wealth of life in the canopy means predators abound and the rainforest canopy supports both the smallest and some of the largest hawks in the world. The falconets of Southeast Asia are no bigger than sparrows and act like flycatchers, sallying from dead branches to catch insects and sometimes small birds. The immense monkey-eating eagle *Pithecophaga jefferyi* of the Philippines and the harpy eagle *Harpia harpyja* of South America are exceeded in size only by the great vultures (the condors and the lammergeier) but are physically the most formidable of all the eagles. They prey on large mammals such as monkeys, possums, and flying lemurs, and they even take small pigs. The monkey-eating eagle is near to extinction; it is now found only on the island of Mindanao, having been extirpated from the other islands of the Philippines. •

A harpy eagle Harpia harpyja *pounces on a howler monkey. The fate of the harpy eagle is of especially acute concern because its territorial requirements are so large that it may now be impossible to establish rainforest reserves of sufficient size to support a viable population.*

3 Within the Forest

ADRIAN FORSYTH

Diversity is the hallmark of tropical rainforests. Other habitats, such as mangrove wetlands, are more productive, and other forests, such as the coniferous forests of western North America, contain more biomass and taller timber. But no habitat rivals rainforest for sheer numbers of species. This diversity is most apparent in the plant forms that are unique to wet and warm tropical regions. As many as half of the trees are buttressed, with spreading flanges at the base that may extend several meters vertically and run up to 10 meters (about 30 feet) laterally. The trunks of the larger trees are typically ramrod straight, tapering smoothly and scarcely branching until they reach the canopy.

DEATH TRAPS

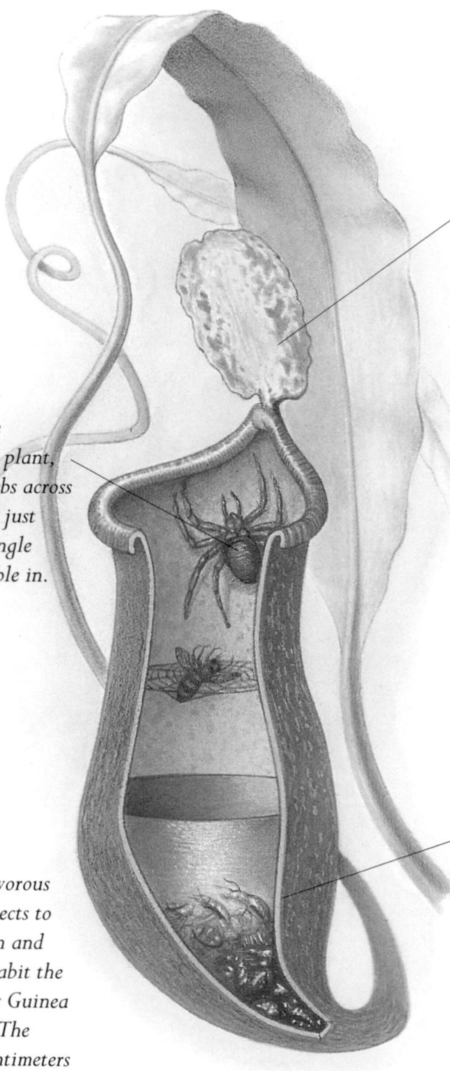

The pitcher is equipped with a lid that functions as a colorful and fragrant lure to insects.

Certain crab spiders take advantage of the pitcher plant, building scaffold-like webs across the inside of the pitcher, just above the water, to entangle unwary insects that tumble in.

Insects that fall into the pitcher are prevented from climbing out because the smooth, waxy inner walls allow them no foothold. They eventually drown in the water, and their bodies are digested by the plant.

Pitcher plants (family Nepenthaceae) are carnivorous plants, which rely on insects to provide essential nitrogen and trace elements. They inhabit the forests of Indonesia, New Guinea and northern Australia. The pitcher may reach 20 centimeters (8 inches) or more in length.

Opposite. *Young Cook's tree-boas* Corallus enydris *shortly after birth. Distant relatives of the infamous anaconda, these snakes live in trees and feed on lizards, birds and small mammals. They are non-venomous, and kill by coiling themselves around their victim, tightly enough to suffocate it.*

The Great Lattice

Strangler figs dangle aerial roots from heights of 60 meters (200 feet) or more, that wrap, intertwine, and self-graft together as they become anchored and consolidated into a sinuous composite trunk, which encloses and kills the host tree. Palms and tree ferns reach gracefully up like gigantic umbrellas. Banyans send down roots from their laterally spreading limbs, creating a spreading phalanx of woody stems that support the ever-spreading tree crown. Lianas, some as thick as ships' cables, contort and twist about each other, while others plaster themselves flat against tree trunks or anchor their claw-like holdfasts in the fissures of the bark. Together, these woody structures form a lattice on which an even richer array of plants flourishes.

The most remarkable tropical plants are the epiphytes, or air plants, which grow on host plants without any rooted connection to the soil of the forest floor. Epiphytes also exist in temperate forests, and even in deserts, but these temperate and arid epiphytes are predominantly mosses and lichens. In the warm, wet regions of the tropics epiphytes include the full array of plant growth forms, from tiny single-celled algae to full-scale trees. Their variety and density are a clear indicator of the relatively benign microclimate of the rainforest interior. This pattern can also be seen on a broader geographic scale. The lowland Asian rainforests of oceanic islands are more continuously humid and generally more profusely endowed with epiphytes than rainforests of the interiors of larger landmasses.

The Water of Life

All plants breathe and move water, as a result of which they are subject to dehydration and cell damage. In temperate or arid climates, freezing and water stress limit the ability of epiphytes and liana-like growth forms to survive. Virtually all temperate orchids and cacti therefore grow in the soil, which allows the roots to tap a relatively huge reservoir of moisture. In contrast, the

Michael & Patricia Fogden

Michael & Patricia Fogden

Top. *Unlike many other ants, the ponerine ants generally lack complex caste systems and social structures, and some species are more or less nomadic. Here a member of this group,* Ectatomma tuberculatum, *attacks a fly on a passionflower in Costa Rica.*

Bottom. *Mimics such as this harmless Ctenuchid moth (*Macroneme *species) enjoy a reduced threat from predators because of their resemblance to dangerous species, in this case a wasp. This method of bluff is known as Batesian mimicry. In complex environments, Batesian relationships can give rise to more subtle systems in which a number of species are all protected to a degree by resembling a common, dangerous species. Such mimics are thus, in a sense, mimicking each other rather than the original dangerous model, resulting in a very intricate system known as Mullerian mimicry.*

majority of the 20,000 or so orchid species found in tropical rainforest grow as epiphytes.

A comparison between rainforest and desert cacti reveals the impact of microclimate even more dramatically. Desert cacti grow in the ground and are deep-rooted at their base. They often have round, barrel, or cylindrical forms that minimize the surface to volume ratio and therefore maximize potential water storage capacity relative to evapotranspiration (the return of water vapor to the atmosphere by evaporation and transpiration). They bristle with defensive spines and hairs that reduce surface wind velocity, reflect light, reduce leaf surface temperatures, and diminish the chance of damage by grazers. These adaptations reduce the plants' light-harvesting capacity, but such a capacity need not be great in the open desert habitat. Rainforest cacti, in contrast, are epiphytic and often virtually spineless, with smooth flat elongate leaves that absorb the diffuse reflected light of the rainforest interior. They may sprout roots at any point, or do without them entirely, simply depending on the high ambient humidity and rains to replenish their water stores.

While water is of paramount importance to epiphytes and lianas, a lack of frost is also crucial. Some lianas run as a narrow conduit for as much as 1,000 meters (about 3,500 feet), and the vulnerability of such a system to ice damage may explain why lianas become progressively more abundant closer to the equator. One hectare (2 ½ acres) of tropical forest may contain 1,000 liana stems—up to 20 percent of the total plant biomass—and in some forests, particularly in the Americas, they knit the forest together in a network running from one tree to another.

It is not, then, the tropics in themselves that are conducive to these growth forms. Plant diversity plummets on tropical mountains high enough for wind and frost damage to occur, and lianas and epiphytes disappear. Studies show that tropical plant diversity is strongly correlated to high rainfall that is evenly distributed throughout the year. This is true for the rainforest trees themselves. Some areas of the Amazon, for example, have as many as 283 tree species per hectare whereas temperate or dry forests may have only one or two dozen. Trees are buffered from moisture stress by their connection with soil; and as one might expect, the correlation between climate and diversity is demonstrated even more strongly by the epiphytes.

Predictable warmth and humidity are the key elements. Perhaps the best indicators of high humidity are the tiniest epiphytes, the epiphylls, including the algae, lichens, mosses, and even ferns that grow on the upper surface of leaves. While some epiphytes may draw upon water in the accumulated mats of humus, detritus, and moss on limbs or in tree crotches, the epiphylls have no such opportunity. The fact that epiphylls find enough water on rainforest leaves, which often have waxy cuticles and drip tips that expedite the shedding of rainwater, testifies to the frequency of rain and to the high humidity in the rainforest.

Leaves and Light

Compared to the rainforest canopy, the forest interior is much less stressful physiologically. The canopy receives direct and intense solar radiation, drying and damaging winds, and torrential rainfall, but it shields the interior from such conditions. Temperature fluctuations, wind speeds, and light intensity are less extreme in the understory, and humidity is higher.

The light at the lower levels is extremely diffuse, and there is little direct sunlight except in the form of sun flecks that constantly shift position as the sun moves overhead. There is in fact a light gradient, and large patches of direct sunlight become increasingly common as one nears the canopy. In contrast to the bright green gloss of canopy leaves, shade plants often have a blue-green sheen and pigments that absorb the red wave-

Leo Meier/Weldon Trannies

Lowland rainforest in northeastern Queensland, Australia. In many parts of the tropics, lowland rainforests are more acutely threatened than highland forests because, being more level and accessible, they are easier and less costly to log, and the land on which they stand is more heavily in demand for various agricultural crops such as sugar cane.

lengths used in photosynthesis. The low-light interior of the rainforest is responsible for the evolution of many of the beautifully colored ornamental plants suitable for cultivation in the mild atmosphere of human households.

The effect of the changing vertical distribution of light availability can often be seen in the way some plants grow. Climbing plants of the Araceae family, such as *Monstera* of Central America, grow upwards from the forest floor. The seedlings orient towards dark shapes that generally indicate a tree trunk, and on reaching one they produce flat shingle-like leaves that plaster themselves vertically and securely against the bark surface. These vertical leaves are well-positioned to receive diffuse reflected light, and they anchor the vine to the supporting tree trunk. When the vine reaches the upper, more open areas of the forest interior and is exposed to more direct light it changes its leaf shape radically. Instead of flat vertical leaves only a few centimeters wide it produces large horizontal spreading leaves that are held away from the trunk in a posture designed to intercept direct sunlight. Trees of the rainforest often change their leaf shape as well: they begin as saplings with large leaves; as the plants grow up toward the canopy and become more exposed to direct light and wind, the leaves are smaller and sometimes more complexly margined (that is, rather than being ovals with a smooth curving edge they have complex indentations).

The canopy is often dominated by epiphyte families such as the Ericaceae (including blueberries and rhododendrons), which have small, tough waxy leaves. In the forest interior, however, epiphytes are able to assume the lush and flamboyant architecture that is the hallmark of most tropical vegetation. Epiphytic philodendrons, anthuriums, orchids, and ferns may sport massive broad-spreading leaves 1 meter (3 feet) or more in length. The three-dimensional lattice of the tropical rainforest is both aesthetically ornamented and ecologically enriched by these epiphytes.

The direct contribution of epiphytes to tropical species diversity is enormous. There are about 30,000 epiphyte species spread through 83 families. Orchids alone have 20,000 epiphytic species, while the

Many butterflies sip nectar from flowers, but some other species feed on animal feces, like these two glasswing butterflies feeding at a bird dropping in a Costa Rican forest clearing.

bromeliads have 2,000. A single tree in Venezuela has been recorded as supporting 47 orchid species. When one considers the range of additional rainforest plant families it is possible that a single tree may support more than 100 other plant species. The contribution of epiphytes to the total biomass is also great, and in tropical cloud forest, where they reach their greatest abundance, they can account for up to 40 percent of the leaf biomass.

Diversity Begets Diversity

The climatic conditions that promote the rainforests' diverse plant communities also amplify the diversity found in associated animal communities. Consider the effect of adding a single plant species—say, a vanilla orchid—to an ecosystem. Unlike temperate-zone orchids, vanilla orchids grow as vines. They cling to tree trunks and dangle in space. Their creamy white flowers attract highly specific orchid bees, a family of pollinators which alone includes many hundreds of species. The pollinated flowers produce long bean-like pods packed with tiny seeds and scented with the complex vanilla fragrance. The ripe pods attract fruit-eating bats, which feed on them and then excrete and disperse the seeds. The leaves of the orchid are eaten by leaf-mining fly larvae, and its roots form symbiotic associations with fungi. In other words, the vanilla orchid provides resources for many animals, and the bees, flies, and bats that feed on it participate, in turn, in a wide network of further ecological interactions. It is through linkages such as these that plants act as ecological multipliers. Each species is a unique molecular and ecological entity that affects a broader community of organisms. Some ecologists believe that for every rainforest plant species an additional 20 animal species occur in an ecological community. Most of these are insects that graze on the foliage and are in turn attacked by arthropod predators and parasites.

The unique architecture of tropical plants provides other ecological resources for the rainforest animal community. *Heliconia* plants typify this: plants in this genus are large herbs of the banana plant family, the Musaceae, that grow as high as 10 meters (about 30 feet) and are characterized by long broad leaves and a gigantic flower-bearing stalk.

A number of rainforest animals have evolved to exploit and depend on aspects of these plants. Some creatures, such as frogs, katydids, and earwigs, find shelter in its leaves, which provide a humid protective cone as they unfurl. Tent and disk-winged bats bite through the mid-rib of mature leaves, collapsing them into shelters that protect them from predators, rain, and sunlight. In many cases the flower-bearing stalks of *Heliconia* grow as erect columns, and the flowers emerge out of

heavy water-filled bracts. The specialized tubular flowers attract hummingbirds, which feed on the copious nectar supply and act as selective long-range dispersers of the pollen. The flowers also house mites that feed on pollen and move from clump to clump by riding on the hummingbirds' bills. The heavy bracts and water are thought to protect the flowers from thieving attacks by stingless bees, which often gnaw through the base of flowers and extract the nectar but move no pollen. In these bracts live small aquatic communities of zooplankton, bacteria, mosquitos, and fly larvae, some of which are found only in *Heliconia* bracts. It seems clear that the specialized flower architecture and interactions between the plant and pollinator have generated a series of ecological and evolutionary opportunities for other organisms. This multiplier effect of plant–animal interactions is described in the ecological aphorism: diversity begets diversity.

It is not only epiphytic plants that make extensive use of the elevated forest lattice. Creatures such as frogs exhibit a comparable upward radiation in tropical rainforest. Most temperate-zone frogs are confined to ponds, rivers, and forest understory habitats. In part this is because their moist gelatinous eggs lack shells and require abundant moisture. In addition, the frogs' highly permeable skin is used in respiration, and the frog is subject to dehydration. The impact of evapotranspiration is also highly significant for cold-blooded animals such as frogs. Evaporation takes place most rapidly under conditions of low humidity,

Some species of glass frogs (family Centrolenidae) have skin so transparent that their bones, muscles, and internal organs can be seen through it. Inhabiting American rainforests, they lay their eggs in clusters on leaves overhanging quiet streams or rivers. In this species, Centrolenella valerioi, *the female abandons her eggs, leaving them to be guarded by the male until the young hatch and drop into the water below.*

and the process extracts heat energy from the frogs' evaporative surface. At about 20°C (68°F) a frog will lose 20 percent of its metabolic heat energy if the air is at 100 percent humidity. But if the humidity drops just a little, to 94 percent, the heat loss in the atmosphere can equal all the heat generated by a frog's metabolism. Consequently, by minimizing evaporative heat and water loss, the humid rainforest interior allows frogs a great degree of physiological freedom.

Rainforest frog communities are species-rich, and the frogs are characterized by a wide range of breeding habits. For example, while there are only three frog species in all of Great Britain, one site in Ecuador supports 81 species of breeding frogs. And whereas temperate-zone frogs lay their eggs in ponds, lakes, and rivers, the majority of rainforest frogs (including 49 of the 81 Ecuadorian species) lay theirs either terrestrially or in the vegetation. The advantage of this for frogs is that it removes the vulnerable egg stage from the reaches of tropical fish, shrimps, and aquatic insect fauna, which prey heavily on frog eggs when given the opportunity. In fact river systems in tropical rainforests are as poor in frogs as the trees are rich in them. It is not the wetlands of the tropics but rather the rainforest itself where the frogs have radiated.

Much of this diversification is focused on choice of breeding site. A common reproductive tactic used by rainforest tree frogs in the genera *Hyla*, *Agalychnis*, *Centrolenella*, and *Phyllomedusa* is to deposit eggs on leaves overhanging water. When the tadpoles hatch they wriggle down into the water. It is also common for male *Centrolenella* frogs to defend the eggs from insect predators as they develop on the leaves. Many rainforest frogs, especially in the genus *Eleutherodactylus*, practice direct development in which no free-living-tadpole stage occurs. Instead, eggs are laid in humid locations, and the froglets, which develop entirely within the egg, are mobile as soon as they hatch. Poison arrow frogs in the genus *Dendrobates* often carry the tadpoles on their backs or deposit them in water-filled tank bromeliads.

A long-tailed hermit Phaethornis superciliosus *takes nectar from a heliconia flower. Many hummingbirds are clad in glittering metallic shades, but the hermits are a group of hummingbirds so-called because of their sober plumage and quiet, retiring habits. Most species live in the undergrowth and lower stories of the forest.*

Michael & Patricia Fogden

Biodiversity in the Forest

It is interesting that tropical forests that are less structurally complex than mature rainforests show much less biodiversity. Asian mangrove forests, for example, are among the most productive habitats on Earth in terms of the sheer volume of biomass that is generated photosynthetically and manufactured from their rich alluvial sediments. Yet mangrove forest typically supports only a few primate species, such as proboscis monkeys in Borneo, whereas adjacent rainforest will support a far richer population that may include tarsiers, slow loris, orang-utans, and gibbons as well as leaf-eating monkeys and macaques. This is due both to the greater variety of food and to the structural complexity of a mature rainforest.

In Asian rainforest the primates tend to occupy particular layers of the forest, using a particular spectrum of resources. Orang-utans, the heaviest primates in these forests, occupy the mid-strata where they feed on foliage, fruit, and strip bark. More omnivorous macaque species spend more time lower down, and the highly agile gibbons and leaf-eating monkeys occupy the upper reaches of the forest where photosynthetic activity and the bulk of the foliage are concentrated. Rainforest in Central and South America has comparable patterns of stratification, although leaf-eating howlers and fruit-eating spider monkeys spend more time in the canopy, and the highly insectivorous species such as squirrel monkeys forage at all levels, even in the understory.

There are additional physical factors that may favor certain plant and animal ecological strategies according to their position in the vertical space. As many as half the canopy trees and vines may depend on the wind or a combination of exploding seed capsules and gravity to disperse their seeds. Canopy epiphytes such as the bromeliads and orchids are primarily wind dispersed.

But these options are less effective for plants of the forest interior; between 70 and 90 percent of these rely on animals as seed-dispersal agents. The most obvious manifestation of this phenomenon is the abundance and variety of fruits and berries found in a tropical rainforest.

A Diet of Fruit

Ecologists conventionally define a frugivore as an animal whose diet is at least 50 percent fruit. As much as 80 percent of the biomass of Amazonian forest is composed of frugivores, including birds, bats, monkeys, rodents, some carnivores, and even specialized fish.

Birds are the most important dispersal agents in the rainforest. This is especially true for the canopy but it also applies to the forest interior. Only birds and monkeys have well-developed color vision, and the conspicuous presence of colored fruit is a measure of their impact. Many of the bird-dispersed fruits consist of either berries, which are packed with small seeds that are swallowed and passed through the digestive system, or larger seeded varieties that are coated by a thin and often brightly colored oil-rich coat known as an aril.

Mammals—bats in particular and then monkeys—are also important dispersal agents, especially for large-sized fruits. Even carnivores eat large quantities of fruit. In the New World tropics the kinkajou *Potus flavus*, a prehensile-tailed relative of raccoons, is primarily frugivorous. In the Old World tropics civets and other viverrids eat much fruit. Large terrestrial mammals are relatively more important in Asian and African forests than in South America.

The size and packaging of a seed is often correlated with its place in the forest. Strangler figs, for example, must be dispersed into the crown of another canopy tree where there is sufficient accumulated debris to support the young fig seedling until its dangling aerial roots contact the ground. These figs use a scatter-gun approach, producing many small seeds in a relatively sweet fruit that is widely sought by a great variety of animals including (in Asia) orang-utans and other primates, small mammals such as squirrels and bats, and a great variety of birds including hornbills, pigeons, and parrots. New World figs are equally favored, and many ecologists believe that fig trees, fruiting as they do throughout the year, play a crucial role in sustaining Asian and New World wildlife populations.

Sweet, small-seeded fruits are often pioneer plants—the early successional species that colonize landslides, new alluvial deposits, and clearings. They do this effectively because of the abundance and variety of the dispersers they attract.

Scientists studying the *Cecropia,* a New World genus in the fig family, found its fruits were eaten by 8 species of monkeys, 12 species of bats, and 76 species of birds. Of course it is always difficult to determine which species are effective seed dispersers. Some birds including parrots and doves, along with primates such as

Over eons, flowering plants have evolved many ways of increasing the likelihood that their animal pollinators will next visit a flower only of their own species. A Madagascar orchid and its hawkmoth pollinator illustrates the extremes sometimes reached in this intricate process: the hawkmoth has a tongue exceeding 25 centimeters (10 inches) in length, and only it can penetrate the depths of the orchid's flower.

Jean-Paul Ferrero/AUSCAPE

A crab-eating macaque Macaca fascicularis *dining on figs. A native of Borneo and Sumatra, this macaque does indeed eat crabs on occasion, but it is on the whole an arboreal species, leaving the forest floor to its larger and heavier relative the pig-tailed macaque.*

orang-utans, act as seed predators by attacking the fruit while it is unripe or by crushing and digesting the seeds.

In contrast to the small-seed strategy, many rainforest trees produce large seeds that attract more exclusive guilds of dispersers. Trees with large single seeds enclosed in a fleshy coat include the avocado family (Lauraceae), the Bursaceae, the Myrtaceae, and the palms. Certain large and durable seeds such as the palms' are gathered, carried away from the parental tree, and hoarded in piles by squirrels and agoutis. Those that go undiscovered and uneaten provide the new generation of palm seedlings. The most specialized of these large-seeded dispersal systems involve a guild of fruit-eating birds that includes hornbills and many pigeons in the Old World, and toucans, quetzals, bellbirds, and their relatives in the New World. All these birds have a relatively large gape that enables them to swallow the fruit whole. Their stomach has a reduced gizzard which removes the flesh without damaging the seed, and the seed is then regurgitated and expelled.

The flesh of many of these fruits is highly nutritious. Lauraceous fruits such as the avocado contain 34 percent oil and 20 percent protein, and some birds, including quetzals, rely on the fruit of the Lauraceae family for four-fifths of their food. Although they may feed on nestling insects, lizards, and animal protein, much of their diet is made up of regurgitated pulp, and it is thought that successful breeding cannot occur without these trees. Because the trees have such heavy seeds it is also difficult for dispersal to occur without the birds.

Some birds such as oil birds and bellbirds are able to subsist and raise nestlings entirely on a diet of fruit. Fruit is typically so abundant that many of these highly specialized frugivores can gather adequate food without male assistance. The female raises the nestlings alone. The males in turn have evolved elaborate plumage and courtship rituals as a means of attracting additional females and increasing their reproductive success.

Toxic Seeds

Many large seeds have defenses against seed predators. In some cases seeds designed for gut passage by large mammals will have a sturdy seed coat, and bird-dispersed large-seeded fruits will often have seeds that are chemically defended. The fruit of nutmegs, for example, split open to reveal a large seed covered with an orange-colored aril. This seed is also impregnated with highly aromatic compounds that in nature act as a warning and a toxic deterrent. The large pigeons of the area swallow the seed and aril whole, but the seed is regurgitated once the aril has been removed. Ground, and used in minute amounts, the seeds are used by humans as a spice, but humans who ingest a single nutmeg seed suffer hallucinations and may enter a coma. The effect on a much smaller mammal such as a squirrel would be far more profound.

The cashew of commerce, originally native to Amazonia, is another example. Its juicy and sweet stem is a swollen, bright orange fleshy affair sought by animals such as spider monkeys. The actual seed, the cashew nut, is located at the end of the stem, and has a seed coat that is impregnated with toxic oils; the monkeys eat the fleshy stem and discard the seed.

Many rainforest trees, especially those of the forest interior, attract mammals such as bats and monkeys. These trees often produce large cauliflorous fruits (fruits that grow directly out of the surface of the main trunk and limbs of a tree rather than at the tips of branches). The resulting solidity enables the tree to support large conspicuous fruits that attract animals with large appetites. The cacao pods, which develop on the trunks and main limbs of *Theobroma* trees, resemble elongate, smoothly ribbed melons. They turn yellow-orange, a color reputedly attractive to New World primates. Monkeys break and gnaw the pods open to uncover a typical "suck-and-spit" fruit, an assemblage of sweet white outer pulp surrounding an inedible seed. Like the nutmeg seed the cocoa seed is laced with bitter defensive compounds, which humans have adapted, in this case to make chocolate. Many of these types of suck-and-spit fruits—including rambutans, mangosteen, lychees, inga beans, and durians—are cultivated and sold in tropical markets.

Perhaps the most spectacular of all cauliflorous fruits are the jackfruit (genus *Arctocarpus)* of Southeast Asia, some species of which reach lengths of 1 meter (3 feet) and weigh up to 50 kilograms (110 pounds). The trees that produce them are not large and rarely if ever reach canopy heights. Ripe jackfruit have a strong odor that is typical of fruits dispersed by nocturnal mammals, and they attract the giant "flying fox" fruit bats in particular. Instead of feeding in the tree crown the bat removes as much as 200 grams (7 ounces) of fruit and carries it away to a roost–feeding site, thus depositing the seeds far from the parental tree.

The forest trees that germinate on the forest floor, in the dark confines of mature rainforest, typically have large seeds, and the fleshy coat that encloses them contains the food reserves needed by a seedling until it reaches an area of increased light intensity.

The Variable Forest

Many rainforest plants are adapted to germination conditions of high direct sunlight. This is true of canopy epiphytes as well as terrestrial herbs and trees. The canopy and the forest in general are broken by light gaps. When a tree falls, either because of wind or death due to disease or insect attack, it often demolishes other trees in its path, and the interlaced lianas may tear branches from neighboring trees or pull them down. The result is a shaft that allows a large area of sunlight to reach the forest interior. A tropical forest therefore is not a homogeneous stand of timber with a monolayer of canopy; tree crowns occur at every level and only rarely in discrete, discernible layers.

Light gaps are rapidly occupied by fast-growing, weedy pioneer plants such as the Melastomaceae, Solanaceae (nightshades), Piperaceae (pipers), Zingiberaceae (gingers), and Musaceae (bananas) families, as well as aggressive climbing lianas and rattan palms. The same fast-growing trees that occupy large light gaps create secondary forests after logging, fire, and other major disturbances. Often they have light

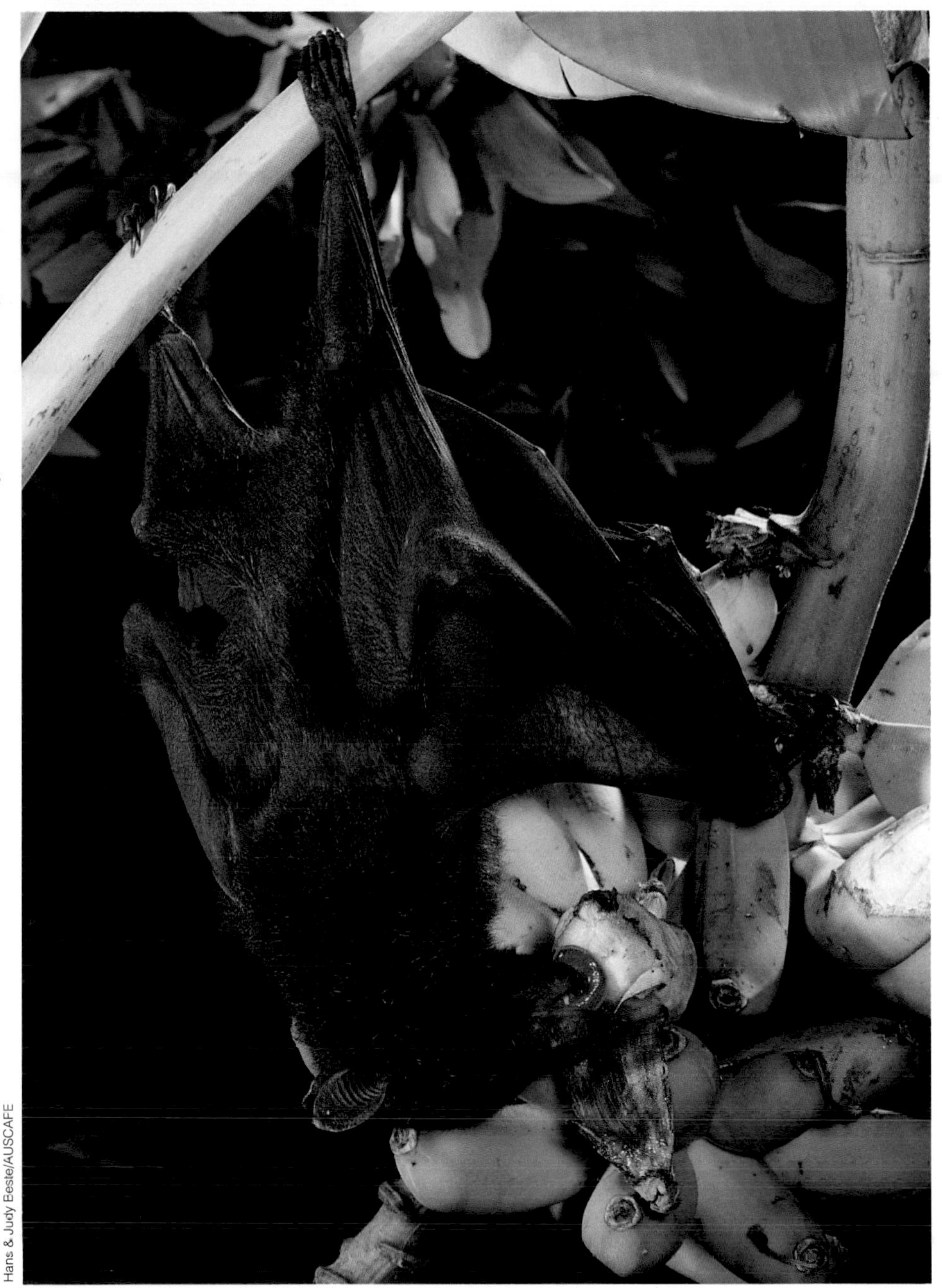

Hans & Judy Beste/AUSCAFE

Black fruit bats Pteropus alecto *roost by day in dense camps (sometimes containing more than 100,000 individuals), which are often located on mangrove islands in the estuaries of tropical rivers. By night they scatter, often flying considerable distances, to feed at a variety of fruiting and flowering rainforest trees.*

Some rainforest animals, including snakes, lizards, geckos, frogs and squirrels, have developed gliding mechanisms, which help them move from tree to tree. Even small gliding surfaces serve to significantly lengthen the span of a jump, promote maneuverability while in the air, or slow the rate of a fall.

HIGH FLYERS

paradise flying snake
Chrysopelea pelias

flying dragon
Draco volans

flying gecko
Ptychozoon kuhli

Wallace's flying frog
Rhacophorus nigropalmatus

Australian sugar glider
Petaurus breviceps

white wood, which is free from the colored resins that defend slower harder trees against insect and fungal attack. Some plant seeds will only germinate when they are exposed to direct light, and opening up the forest stimulates a distinct set of seeds to become active. Many light-gap plants flower and fruit profusely within a brief time. But their leaves are less well-defended chemically than mature forest understory or canopy leaves, and they attract large herbivorous insect populations. These factors also attract heat-loving lizards and snakes. Certain insectivorous birds forage in light gaps as do their nocturnal counterparts, the bats.

Light gaps are short-lived. Taller woody trees are hardier than pioneers and eventually the gap is filled by the shady crowns of its more dominant occupants. The rate at which trees fall and gaps are created and filled determines much of the character of the forest interior. Forests subject to high winds and cyclones are more dynamic, heterogeneous, and shorter. Soil conditions, waterlogging, and slope also influence this dynamism. The presence of a river system may cause sudden changes in the forest: forest lands may erode rapidly on one side of the river and rich bare alluvium may be laid on the other; the river may change course, leaving shallow lakes to be filled in with swamp vegetation.

The classic rainforest, which is tall and diverse with an open interior, typically occurs on relatively flat well-drained soils of moderate to low fertility, in areas that are not subject to severe winds or fire-inducing droughts. In general, increases in elevation, poor drainage, and high soil acidity result in a forest of reduced height and species diversity that has a claustrophobic interior—filled by stems and trunks of smaller average size. Heath (or white-sands) forests grow on relatively sterile, coarse quartz soils and are short and composed of trees of small diameter and uniform height. The heath forest appears as a dense stand of poles capped by a dense canopy, and these low-diversity formations support far fewer numbers and kinds of animals. For example, heath forest in Sarawak is reported as supporting only 24 frog species while lowland rainforest supports up to 53. The numbers for lizards and snakes are similar: 13 heath lizard species compared with 31 in the rainforest, and 13 heath snake species as opposed to the rainforest's 35—that is, compared with a typical rainforest, heath forest supports less than half the diversity of reptiles and amphibians.

Rainforest formations differ considerably within a country, according to various physical factors, and they also appear to differ on a much broader geographical and historical scale. The bromeliads, for instance, that are so abundant in the New World rainforests are absent from the Old World tropics. The insectivorous pitcher plants (genus *Nepenthes*) that are so important in many Southeast Asian rainforests have no equivalent in Central America or Africa. Impressively large terrestrial mammals in New World rainforest consist only of tapirs, pigs, jaguars, and at higher elevations, spectacled bears. In contrast Asian rainforests boast rhinoceroses, tigers, banteng, pigs, bears, and elephants, whereas African rainforests are richer still in large grazing understory mammals. The arboreal mammal fauna also differs from one continent to another. The sloths of the New World have no Old World counterpart. Sap-sucking monkeys are not uncommon in South America but are unreported elsewhere. The gigantic fruit bats of Southeast Asia have no New World equivalent.

THE FIFTH LIMB

Mammals and reptiles of many kinds have evolved prehensile tails as an aid to locomotion in the trees. Aside from the added security against falling, such an aid also has the potentional advantage of freeing the hands for delicate manipulation and for obtaining food items otherwise unreachable. Pictured from left to right are the prehensile-tailed porcupine, the tamandua, and the kinkajou, all inhabitants of tropical American forests.

Locomotion Strategies

Differences in rainforest structure on different continents are thought to be responsible for some of the different locomotion strategies that have devleoped in arboreal rainforest animals. For example, it is extraordinary that prehensile tails have evolved so many times in the New World forests when gliding has evolved so often in the forests of Southeast Asia.

A prehensile tail is by definition capable of supporting its owner's full body weight. It can function as a fifth limb, and indeed most prehensile tails weigh as much as a limb and require a substantial rewiring of the nervous circuitry used to coordinate locomotion. It is therefore remarkable that this adaptation has evolved in many opossums, several genera of monkeys, the kinkajou, at least two species of anteater, porcupines, and several pit viper snakes. All of these animals make extensive use of their tails to anchor their bodies while moving and feeding above ground. In contrast, only pangolins in Southeast Asia rely on a strong prehensile tail, presumably to anchor themselves when ripping open ant and termite nests.

Asian animals distinguish themselves by their reliance on gliding. Many animals ascend the tree trunks and then glide either to the ground or to an adjacent tree. This behavior has been developed by seven genera of squirrel, the flying lemurs, two geckoes, two *Draco* lizards, some frogs, and even a snake in the genus *Chrysopelea*. These animals are all true gliders in the sense that they possess adaptations that increase their surface area, and they employ a flattened concave ventral profile when they leap from their perch. This habit is virtually absent in the New World, and neither gliders nor animals with prehensile tails are common in Africa. Some biologists suggest that this reflects the fact that lianas, the lateral highways of the forest interior, are relatively abundant in New World rainforests but uncommon in the taller Asian forests, Africa is intermediate in character. Other biologists are sceptical of such generalizations. Perhaps it is enough to recognize and enjoy the uniqueness that every rainforest offers. ■

VINES AND STRANGLERS

FRANCIS E. PUTZ

The major threat to most young rainforest trees is shade—being overtopped by neighbors is tantamount to death, or at least suppression. As little as 1 percent of full sunlight reaches the forest floor beneath a full canopy, and tree seedlings can only grow quickly in the vicinity of light gaps. But increased height must be supported by increased girth to ensure stability, and most of the mass of a large tree is invested in structural material, namely wood. If a trunk is not sound the tree will probably fall or die in the shade.

Many plant species of the canopy circumvent the need to be self-supporting and avoid the harsh conditions of the forest understory by relying on trees for support. Some of these plants climb upwards from the forest floor by grasping onto trees. They are called vines, climbers, bush ropes, or, if woody, lianas. Others, which start in the canopy and grow down to the ground, are referred to as hemiepiphytes because they spend part of their lives as epiphytes but later become rooted in the ground.

Strangler figs start out as seedlings on a branch, growing slowly downward until a root reaches the ground, after which growth accelerates.

There are at least 2,500 species of vines in the world, from about 90 plant families, including climbing ferns, climbing gymnosperms, climbing bamboos, and climbing palms. Vine seedlings on the floor of a tropical forest are faced with the extremely difficult task of climbing up to the canopy, often 30 to 60 meters (100 to 200 feet) overhead. They are generally self-supporting to a height of about 1 meter (3 feet), after which they can only grow higher by attaching themselves to a suitable trellis.

A vine's specific requirements for support depend on the way it climbs, and vines have evolved a wide assortment of mechanisms for grabbing onto, hooking into, sprawling over, or otherwise clinging to trellises. Many vines cling to trellises with coiling tendrils and in others the stem itself twines around supports. Only vine species with adhesive tendrils or roots specialized for clasping bark can climb up the trunks of large trees.

One root follows another, the roots fuse, and a woody cylinder is formed around the supporting tree

Vines that twine are generally limited to climbing trees less than 20 centimeters (8 inches) in diameter; on larger trees the coils slip down and the vine is dislodged. For those few vines that encounter a series of successively taller supports of the appropriate diameter, ascent to the sunlit canopy entails passing from a tree seedling to a sapling, from the top of a sapling to the lower branches of a taller tree, and so on to the top of the forest.

Once in the canopy, vines generally spread from tree crown to tree crown, displaying their foliage amidst the leaves of their supporting trees. Surprisingly, given that vine stems are generally quite slender, their leaves often comprise 20 to 40 percent of all the leaves in a rainforest canopy. This is possible because of the distinctive "plumbing" in most vines. Despite their narrow stems, vines have water-conducting vessels that are huge in comparison with those in the wood of most trees. Because flow rates increase rapidly with vessel diameter, many leaves can be sustained on the stems. Vines can make the most of this advantage because, unlike trees, they do not provide support for their branches and leaves.

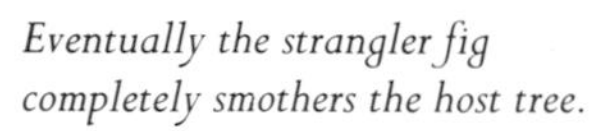

Eventually the strangler fig completely smothers the host tree.

Hemiepiphytes, such as strangler figs, bypass the problem of reaching up to the canopy but face the equally difficult problem of getting down to the ground. By starting as seedlings perched in tree crowns they avoid the dense shade of the forest understory, but they depend on atmospheric inputs of water and nutrients. Even in an extremely wet forest the hot sun and severe water deficits between rainshowers produce desert-like conditions in the canopy and the hemiepiphytes grow extremely slowly. Successful hemiepiphytes overcome this basic problem by growing roots down to the ground where water supplies are more dependable. No one knows how long it takes a hemiepiphyte seedling to become rooted in the ground, but given that epiphytes suffer from almost daily water stress, the time must be reckoned in decades. After hemiepiphyte roots reach the ground, growth rates increase and the plant changes form in response to increased access to water. In the epiphytic phase, hemiepiphyte leaves display many of the characteristics of desert plants; they are thick with water-storage tissue and their few stomates (breathing pores) are closed much of the time to reduce water loss. The trade-offs of this water-conserving strategy are reduced rates of carbon dioxide uptake, reduced photosynthetic rates, and slow growth. Ground-rooted hemiepiphytes dispense with these water-saving strategies and grow rapidly.

The growth patterns of vines and hemiepiphytes represent radically different evolutionary "solutions" to the shared problem of avoiding the dense shade of the tropical forest understory. ●

Leo Meier/Weldon Trannies

Michael & Patricia Fogden

*Detail of a tree trunk in a Costa Rican rainforest with an aroid (*Monstera *species) and numerous other plants climbing up it.*

Left. *One of the characteristic features of tropical rainforests is the abundance of lianas which often form incredibly tortuous tangles and may themselves serve as supports for other plants such as mosses, ferns and orchids.*

Below. *In most plants, a cross-section of the stem shows a more or less cylindrical arrangement of tissues, which gives physical support and facilitates the transport of nutrients to and from the plant's extremities. But a cross-section of the stem of many lianas, such as those illustrated below, reveals a more complex arrangement the woody internal support tissues grow irregularly to form stranded, rope-like structures that provide strength and flexibility no matter what the orientation of the stem.*

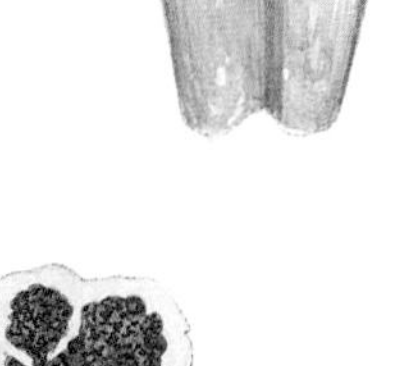

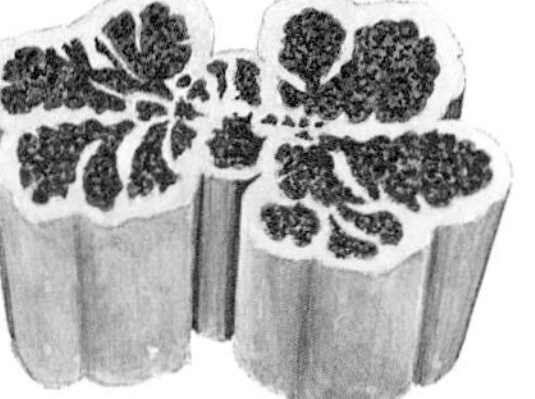

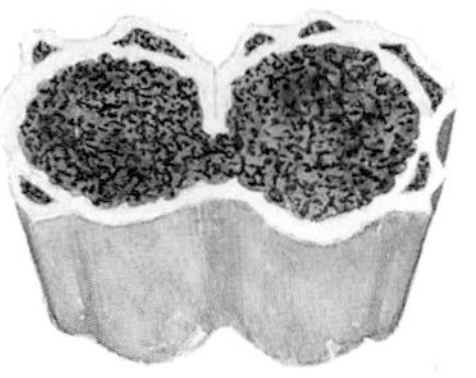

4 LIFE IN THE SHADE

FRANCIS H.J. CROME

Overhead there is endless green; below, rich brown humus and an occasional mute rotting log; on all sides, phalanxes of tree trunks and saplings. This is the cool, quiet, dark world of the forest floor, where a Byzantine assortment of life forms—fungi, worms, insects, elephants, giant flowers, leeches, and shrews—make their homes far from the life-giving light.

THE GREEN MANSION

The rainforest has been likened to a cathedral with a soaring vault of trees covering and protecting a calm shady gloom where humans may walk more or less unobstructed between the trunks of giant trees. Such places do exist and popularly typify "real" virgin rainforest, but they are less common than one might think. They only occur in certain types of unlogged rainforest, where the physical conditions are optimum and where there has been no disturbance to the canopy for long periods. More often the floor presents a patchwork of conditions that are the legacy of the forest's history—the patterns and timing of treefalls, landslides, fire, storms, and human-induced disturbances. Recently formed and brilliantly lit gaps with their tangles of vines, bamboos, rattans, and saplings are interspersed with older areas where the canopy has covered over but dense stands of saplings still remain, and with the very oldest areas—undisturbed for decades or centuries—where the understory has thinned out to produce the classic green mansion.

Opposite. *Notable for their technicolor faces, the mandrills of Africa are rainforest relatives of the baboons.*

A star stinkhorn fungus Aseroe rubra *in cloud forest, Costa Rica. Stinkhorns disperse their spores through the agency of blowflies, which are attracted by their noxious smell.*

Michael & Patricia Fogden

WAITING FOR LIGHT

Where the canopy is complete only between 5 and less than 1 percent of sunlight reaches the forest floor. That light is also changed in quality from full sunlight, having a higher proportion of longer-wavelength far red and infra-red light. Seeds that fall to the floor and the resultant seedlings encounter special problems in these light conditions. For example, small seeds such as those of pioneer trees usually require high light to germinate and grow.

Some species of primary or secondary trees have large seeds that provide sufficient food reserves for the seedling to reach 1 meter (3 feet) or more in height. Germination time can vary immensely from a few days to years: the 5-centimeter (2-inch) diameter seeds of the yellow walnut *Beilschmiedia bancroftii* of north Queensland can remain viable for up to 11 years. Those that do not germinate, rot, or get eaten, form a seed bank in the soil and leaf litter that plays a vital role in forest regeneration.

Once they have used up their seed reserves, most seedlings need more light to grow than is found a meter above the forest floor. Saplings of many if not most species may have to sit in the dark for years or decades, showing no appreciable growth while they wait for a light gap overhead to provide the conditions necessary to grow and push up into the canopy. In dipterocarp forest in Southeast Asia saplings 5 centimeters (2 inches) in diameter increase in thickness by less than 0.7 millimeters (1/32 inch) a year and in some Australian forests saplings of many species may even shrink over periods of several years. Other trees can grow, albeit slowly, by taking advantage of the sun flecks that cross the forest floor. They capture this fleeting energy in special pigments and later mobilize it to construct the sugars that normal photosynthesis provides. Many species in the shade have red undersurfaces beneath the photosynthetic tissue that are thought to improve photosynthesis by back-scattering light through the photosynthetic tissue.

The leaves of the saplings on the forest floor differ substantially from those in the canopy, even in the same species or on the same tree. The canopy leaves

are leathery and adapted to conserve moisture in the open sunlight, while in the shade they are much larger, loose, lax, and soft. In forest at Amani, Tanzania, shade leaves of *Myrianthus arborea* are 8 times larger than sun leaves; of *Anthocleista orientalis* 28 times larger. Once the leaves have toughened up they can last for three years or more, and under the right conditions of high humidity and intermediate light levels they may become covered with a garden of lichens, algae, bacteria, and liverworts, some of which fix nitrogen (that is, convert gaseous nitrogen into a soluble form that plants can use).

Michael & Patricia Fogden

Shrubs and Herbs of the Rainforest Floor

A true shrub or herb story is unusual in rainforests. The big-leaved plants and giant herbs such as bananas, which popular belief and old prints would lead us to believe cover the forest floor, only occur under good light conditions. Such leafy luxuriance so low down is only found on forest edges and in gaps large enough for light to reach right to the ground. There are of course shrubs and herbs on the forest floor, but numerically they usually play second fiddle to the welter of saplings of canopy and understory trees.

Michael & Patricia Fogden

Some species of crabs, usually dwellers of the ocean or its shores, have found a place in the teeming diversity of tropical rainforests. The land crab Gecarcinus lagostoma *of the Caribbean coast of Central America returns to the sea only annually to breed, often traveling many miles to do so.*

SUBTLE AND SHOWY BLOOMS

In South America the understory or ground layer has many species of shrubs or small trees, particularly in the families Rubiaceae and Melastomataceae. All produce berries and some have very showy flowers or bright red bracts. In Africa, Asia, and Australia shrubs are less obvious but some form single-species stands. In the Usumbara Mountains of northern Tanzania, two shrubs, *Mimulopsis* species and *Acanthopale laxiflora*, form dense stands that mass flower, the former at three-year intervals and the latter only every 12 years. In some upland forests of north Queensland the shrub *Hodgkinsonia* forms similar stands but it fruits regularly.

Herbs of the forest floor include gingers (family Zingiberaceae), bananas and heliconias (families Musaceae, Strelitziaceae, Heliconiaceae), marantas (family Marantaceae), aroids (family Araceae), ferns, club mosses (genus *Selaginella*), and lilies (family Liliaceae). Such herbs usually have some sort of underground rhizome, can grow several meters high, and often bear spectacular blossoms. The heliconias and strelitzias of South America have complex inflorescences (flower-bearing stalks) consisting of big red or orange canoe-shaped bracts that enclose the tubular flowers. The flowers produce copious nectar and are favorite feeding places for hummingbirds and insects. Water collects in the bracts, and within these

Like these palms and flowering heliconias, shrubs and saplings of the lower levels of tropical forests are often characterized by enormous leaves, which facilitate photosynthesis in the dim light below the canopy.

little microcosms a range of aquatic insect larvae exists. The gingers, which seem able to grow in more closed canopy conditions than the bananas and heliconias, are common on the forest floor in Southeast Asia and Australia. They too produce beautiful flowers, particularly species in the genus *Costus*. The magnificent torch ginger *Nicolaia elatior* from Indonesia forms huge clumps up to 6 meters (20 feet) high, and its inflorescence is formed of brilliant red waxy bracts on stems up to 2 meters (7 feet) high.

However, many herbs of the forest floor are characterized more by the subtlety of their blooms and the beauty of their leaves. The commonly cultivated African violets are shade dwellers from the forests of the Usumbara Mountains in Tanzania. Many aroids, particularly in the genera *Anthurium* and *Allocasia*, have beautiful leaves that are often dark green with white or silver veins. *Anthurium crystallinum* of Colombia has velvety heart-shaped leaves and *A. clarinervium*, which grows in clay in the cloud forests of Chiapas, Mexico, is a dwarf species that has dark velvety leaves with silver veins. *Allocasia lowii grandis* and *A. cuprea* grow from tubers in the forests of Borneo, the former bearing large deep green leaves with white veins and the latter bearing large metallic green leaves that are purple beneath.

PARASITIC PLANTS

Plants of the genus *Rafflesia* are parasites that produce their huge flowers, the largest in the world, on parts of host vines near or on the ground. The flowers of *R. arnoldii* of Sumatra are up to 80 centimeters (about 30 inches) across, thick, fleshy, and have an odor of rotting meat which attracts the flies that pollinate them. The seeds are held within the fleshy body of

Michael & Patricia Fogden

Michael & Patricia Fogden

Like several other rainforest groups, frogs of the tropical American genus Eleutherodactylus *bypass the tadpole stage. Eggs are laid on land, and the embryos complete their development within the transparent eggs, emerging as fully formed froglets.*

species perching on leaves to get a wider coverage of the air currents. Once they have found dung they bury it or form it into balls which they transport to burrows, where it forms the food of the larvae. While some species of dung beetle specialize on monkey feces, others eat no dung at all. Some species make balls of plant material, and *Canthon virens* of Amazonia attack leaf-cutter ants and make balls of the contents of their abdomens.

Fruits and seeds provide one of the most important concentrated resources to both invertebrates and vertebrates, and the species that eat seeds are seed predators. To survive, a seed must escape these predators and many do so by being dispersed by specialized birds, mammals, and fish. These dispersers eat the fruit flesh around the seed and do not destroy the seed. Other seeds escape potential predators by being extremely poisonous.

Some of the toxins developed by plants to counter seed predation are powerful and include chemicals that generate cyanide, latexes, alkaloids such as strychnine, bitter tannins, and poisonous proteins such as ricin in the seeds of the castor oil plant *Ricinus communis*. In turn, insects have developed their own defenses and may even use the poisons themselves. Many legumes, for instance, produce canavanine, which interferes with an animal's ability to manufacture protein. The bruchid beetle, an important seed predator in South America, can not only distinguish the canavanine but uses it as a food. Male butterflies of the family Danaiidae convert the alkaloids in the *Crotalaria* plant to an aphrodisiac they use in courtship.

Forest Floor Vertebrates

TOXIC FROGS

Frogs of the forest floor tend to be nocturnal and cryptic. Many burrow in the soil or hide in the litter and their camouflage can be superb. The horned toad *Megophrys monticola nasuta* of Malaysia, for example, looks just like a dead leaf, and in southeastern Brazil there are two totally unrelated frogs, *Stereocyclops parkeri* and *Proceratophrys appendiculata*, that not only look like dead leaves but when disturbed stretch their back legs out and become totally still for up to 30 minutes, thus increasing their resemblance to a leaf.

Not all frogs have dull colors, however. The poison arrow frogs of the tropical American family Dendrobatidae are active by day, when their reds, greens, blues, and yellows stand out like little jewels among the leaf litter or on the leaves of low plants. Their brilliant colors advertise the fact that they are poisonous, and they are indeed among the most poisonous animals known. Their skins secrete a range of powerful toxins which some native Indians use to poison the tips of their blow darts. In western Columbia the Emberá Chocó and Noanamá Chocó Indians use three frog species in this way.

The most toxic poison arrow frog is the beautiful pure yellow-gold *Phyllobates terribilis* that is reputed to be so poisonous that simply holding one in the hand can cause death. To treat their darts, Indians need only rub the tip over the skin of such a frog, but other species have to be tortured to release enough toxin for the darts.

The skin secretions of virtually all frogs and toads are irritating or toxic, but those of the American poison arrow frogs are among the most lethal substances known, a fact exploited by native Amerindians who use them to poison the tips of their blow darts. Here two male strawberry poison arrow frogs Dendrobates pumilio *wrestle in a territorial dispute.*

Michael Fogden/Bruce Coleman Limited

Michael & Patricia Fogden

Top. *The male hercules beetle* Dynastes hercules *uses his enormous "horns" in battles with other males.*

Bottom. *A three-toed sloth* Bradypus variegatus *defecating in a Panamanian rainforest. Sloths laboriously descend from the tree-tops to defecate about once every eight days.*

SNAKES AND LIZARDS

The most abundant vertebrate predators on the forest floor are probably the reptiles, particularly lizards and snakes. The pythons of the Old World and Australasia and the boas of the New World are the best known, because of the prodigious size they reach, but they are not characteristic. Most snakes of the forest floor are small to medium-sized, cryptic, frequently nocturnal, and poisonous. Terrestrial snakes usually eat a range of amphibians, mammals, small birds, and occasionally insects, but some, such as the South American mussurana, genus *Clelia*, are specialists. The mussurana, a glossy black, rear-fanged snake, often feeds on other poisonous snakes and is able to overpower by envenomation and constriction some of the largest poisonous snakes found in South America. Another genus that specializes in eating other snakes is the triangular-shaped *Mehelya* or file snake from Africa.

In South America there is an extensive mimicry complex involving a group of poisonous coral snakes of the family Elapidae, several harmless snakes of the family Colubridae, and several rear-fanged mildly poisonous snakes also of the family Colubridae. All of these snakes are strikingly similar in appearance, displaying colorful patterns of red and black or red, black, and yellow or white bands. The mildly poisonous colubrids, such as those of the genus *Erythrolamprus*, are believed to be the models in this system, that is they are mimicked by the other two groups.

The largest poisonous snake in the Americas and the second largest venomous snake in the world is the fearsome-looking bushmaster *Lachesis mutus*, which can reach lengths in excess of 3 meters (10 feet). In some parts of its range it is known as "cascabella muta", or mute rattlesnake, for although it looks like the tropical rattlesnake it does not actually possess a "rattle". The bushmaster is a nocturnal feeder, and despite its potentially dangerous venom it is not a serious threat to human beings as it rarely penetrates populated areas. Not so the large species of venomous *Bothrops* snakes, which often clash with humans in much of their range in Central and South America, particularly on plantations. The common barba amarilla *Bothrops atrox*, known to most people as the fer-de-lance (the true fer-de-lance actually occurs only in Martinique), is a veritable land mine—cryptically colored and easily trodden on by unsuspecting animals and often the deliverer of a fatal bite.

Burrowing snakes and lizards, including blind snakes and legless lizards, are also common on the forest floor. Most are insectivorous and fall prey in turn to other burrowing snakes. The mole vipers, genus *Atractaspis*, found throughout Africa, are burrowers known particularly for their disproportionately long fangs. Even when the snake's mouth is closed, the fangs protrude beyond the corners of the mouth and can be used to kill prey—a useful ability when feeding occurs in such restricted areas as rodent burrows. For obvious reasons, this adaptation also makes these snakes one of the more difficult snake species to handle manually.

SPECTACULAR BIRDLIFE

The understory and terrestrial bird population may not be great, but the number of species is. Not only are there specialized terrestrial and understory species, but

Michael Fogden/Bruce Coleman Limited

many canopy species come down to drink or to shelter during hot periods of the day or year. The terrestrial and understory birds, among them pittas, ovenbirds, and babblers, are mostly insectivorous, but there are also several important groups of frugivores and mixed feeders including pheasants, curassows, manakins, and cassowaries. A casual walk through the forest gives an inaccurate impression of the number of species in these lower regions. To get a better picture an observer must use a hide or catch birds in nets.

The true gems of the forest floor are the plump little pittas of the Old World and Australasian forests that hop across the leaf litter rummaging for invertebrates. Although terrestrial, some species ascend high into the canopy to call. Larger species of terrestrial birds are vulnerable to predators, shy, and seldom seen. This is the case with the pheasants of Southeast Asia, the curassows in South America, and the giant crowned pigeons of New Guinea. The Asian forests are the homes of such well-known game birds as the peafowl and the jungle fowl *Gallus gallus*, from which domestic chickens are descended. Less well known and even more rarely seen are spectacular species such as the congo peacock *Afropavo congolensis* of West Africa, the imperial pheasant *Lophura imperialis* of Vietnam, the Palawan peacock-pheasant *Polyplectron emphanum* of the Philippines, and the argus pheasant *Argusianus argus* of Southeast Asia. The argus pheasant performs his courtship displays from a dancing ring he clears in thick forest. He struts about his arena, then will suddenly fan his wings to display a breathtaking curtain of wing coverts that completely hide him and are covered in "eyes". He shivers and rattles his wings as he slowly lowers them down.

Mound-builders are fowl-sized birds named after the huge compost heaps of leaf litter they build for

Cryptic, lethargic, and easily trodden on, the fer-de-lance Bothrops asper *is a rainforest species that is probably responsible for more human fatalities than any other South American snake.*

their eggs instead of nests. The birds work at the mounds to maintain the correct incubation temperature for their eggs. They are found in Indonesia and the Australasian region and include such species as the Australian brush-turkey *Alectura lathami*, the maleo *Macrocephalon maleo* of Sulawesi, and the scrub-fowl or megapode *Megapodius freycinet* of many areas in the Pacific. In some places their building activities actually clear hundreds of square meters of litter from the rainforest floor.

LARGE MAMMALS ARE HARD TO FIND

Large mammals are popularly believed to abound in the rainforest, but again this is a misconception. Most species are small and cryptic, and rodents (including squirrels and porcupines) are probably the most common. In the Old World and Australasia they are often very abundant—in the Australian tropical forest they occur in high densities with trapping rates of 20 percent to 40 percent, which means that if traps were set to catch small rodents alive, this proportion would catch a rodent each night. In tropical America they have radiated into a wide range of forms such as agoutis and pacas. Most are omnivorous and they are important seed dispersers in many tropical forests. The larger rodents are similar in size to the small forest antelopes, the duikers and chevrotains of Africa and Asia.

The pangolins of Africa and Asia and the armadillos of tropical America are insectivores and have similar body armor to protect themselves from predators. Pangolins also have sharp teeth along the plates of the tail, which are useful in that they can be dragged across an attacker, inflicting deep wounds.

Larger rainforest animals, many of which have been persecuted to near extinction, tend to occupy a range of habitats and often prefer more open and disturbed regions where browsing and grazing are easier or more profitable. In Africa these include the forest elephant, the bongo, the okapi, and the pygmy hippopotamus; in

COURTING CHAMBERS

Male bowerbirds build elaborate bowers in which they display to attract females. The bowers are decorated with many small items including feathers, snail shells, small bones, berries, and flowers. Rival males frequently steal display items from each others' bowers.

The satin bowerbird Ptilonorhynchus violaceus *of eastern Australia constructs a typical avenue bower, consisting of two parallel walls with a display area in front.*

An inhabitant of the highland forests of New Guinea, McGregor's bowerbird Amblyornis macgregoriae *builds a typical maypole bower, consisting of twigs arranged around the base of a small sapling.*

The golden bowerbird Prionodura newtoniana *of the highlands in northeastern Australia is the smallest bowerbird but builds the largest bower—a double maypole, with a connecting display perch between.*

D. Parer & E. Parer-Cook/AUSCAPE

An adult male gorilla in a forest in Zaïre. Largest of the living primates, gorillas spend much of their time on the ground, though they can climb easily and often build sleeping nests in trees.

Asia, elephants, Javan and Sumatran rhinoceroses, anoa, tapirs, and deer; and in tropical America, tapirs and deer. All continents support pigs in the forest and they are often major consumers. The bearded pigs of Borneo eat much fruit and go on long mass migrations through the forest. Peccaries are common in America and bush pigs in Africa. The strangest-looking pig is the babirussa of Sulawesi whose upper tusks, which are artfully used in combat with other members of the species, grow through and out of the upper jaw in a tight curve.

In the Old World forests, carnivorous species are commonly civets and genets, and large cats are decidedly uncommon. Leopards and clouded leopards live in pure forest but tigers and jaguars appear to prefer more varied habitats where their large prey is more common. In South America the big cats reach their highest diversity at mid-altitudes but are uncommon in lowland Amazonia.

With the notable exceptions of gorillas, chimpanzees, and mandrills, few primates live on the forest floor. Lowland gorillas are more arboreal than the better-known mountain gorillas, and chimps also spend much time in trees. Mandrills, however, are strictly terrestrial and their bizarre facial ornamentation is thought to be an evolutionary response to the dark conditions in the forest where facial gestures and other signals used for communication are harder to distinguish. ■

LEAK-PROOF NUTRIENT-CYCLING SYSTEMS

CARL F. JORDAN

Large-scale efforts to develop tropical agriculture and forestry in the mold of temperate-zone systems continue, and for the most part, they continue to fail. The reason often lies in the failure of developers to understand how the nutrient cycles of tropical forests function.

When European scientists began exploring the tropical rainforests in the nineteenth century, they were struck by the large size and apparent vigorous and luxurious growth of the trees. Assuming that the soils underlying this frenzy of growth were rich in nutrients, many scientific explorers recommended that the forest be cleared and exploited for agriculture. Time after time, however, crops grew well for a year or two after the forest was cleared, but yield would decline soon after—often drastically.

What caused the decline? Soil scientists discovered that as a result of the high leaching due to the continuous hot and moist conditions, tropical soils are usually very low in nutrients. But if most tropical soils are nutrient-poor, how can a tropical forest maintain a large structure and vigorous growth? That puzzle has intrigued ecologists for almost half a century, and the answer has begun to emerge only during the past few decades.

The tropical forest survives through a system of nutrient recycling in which there is relatively little interchange between the actively cycling nutrients and the underlying mineral soil. It is a system that effectively "short-circuits" the open systems common in temperate agriculture.

How does this so-called "closed" nutrient cycle work? First of all, in contrast to rich temperate-zone ecosystems, where roots penetrate the mineral soil down to 1 meter (3 feet) or more, roots in tropical forest ecosystems generally are concentrated at or near the soil surface. They may even form a mat on top of the soil. Leaf litter and dead wood on the forest floor are attacked by decomposers such as fungi and bacteria, and the nutrients are incorporated in the biomass (bodies) of these microorganisms. Litter may also be consumed by soil arthropods such as termites.

As these organisms go through their life cycle and die, the nutrients are released and can be taken up by other microorganisms. Alternatively, they can be taken up by plant roots, or by mycorrhizal fungi which attach to plant roots and increase their efficiency of nutrient uptake. Because competition for the nutrients is extremely intense, having roots and mycorrhizae right on the soil surface where the litter is being decomposed gives plants an advantage. Nutrients are taken up and recycled by the plants as soon as they are released by decomposers near the soil surface. Thus there is very little opportunity for loss of nutrients. In reality, this cycle is neither completely closed nor "leak-proof". But it is highly leak resistant—as long as the forest is not cut down.

Buttress roots have evolved in many rainforest tree species as an effective means of providing support with a minimum investment in bulk at ground level. These rainforest root systems are so efficient that almost all the nutrients held in decaying plant matter around the roots are cycled back into the living plants, thus creating a lush rainforest on relatively poor soil.

Leo Meier/Weldon Trannies

RAINFOREST

In forest growing on rich soils (above left) roots penetrate deep down and take up nutrients as needed. In poor soils (above right), such as those that occur in much of the tropics, roots are concentrated near or on the soil surface where they intercept and take up nutrients before they are lost at deeper levels. Tiny root tips appear to attach themselves to recently fallen leaves (detail at right), and this ensures few nutrients escape reabsorption.

When tropical rainforest is cleared, the nutrients that are released by burning and decomposition are immediately incorporated by microbes living in the soil. As they are released through excretion or through the organisms' death, the nutrients become available for uptake by crops, trees, or grasses. That is why productivity is often high in the year or two following tropical deforestation.

However, as the litter and other organic matter remaining on the forest floor decompose and disappear, the soil microorganisms no longer have a carbon source for energy and their population crashes. At the same time, because the root network of the crop is small and thin, the system develops leaks. Potassium, which is held only weakly in the soil, is lost through leaching. Nitrogen is quickly volatilized, because there are few microbes to take it up and conserve it. Phosphorus too is lost. When phosphorus occurs in an inorganic form in acid tropical soils, it is quickly immobilized by iron and aluminum. After a few years the system becomes nutrient-deficient and productivity declines. Because hauling fertilizers into rainforest areas is prohibitively expensive, the sites are usually abandoned. We are left with neither a beautiful rainforest nor a productive commercial system.

Given the fragile nature of tropical nutrient cycles, can tropical ecosystems be managed so they allow a sustainable yield without destroying the nutrient capital? This is an important question for tropical countries that desperately need to feed their peoples and to grow crops for export.

Tropical ecologists think that cropping systems that mimic the natural forest as much as possible offer some hope. This means that instead of growing monocultures of annuals such as corn and upland rice, or of pastures that are easily destroyed by cattle, productive systems should have a large component of trees. Preferably, the trees should yield products such as rubber, brazil nuts, and tropical fruits that can be harvested without destroying the structure of the forest. Harvesting such systems removes only a small proportion of the total biomass and the nutrient stocks, and scarcely disturbs the root systems. In addition, any cropping system in rainforest soils should consist of many types of plants, because each species has its own nutrient requirements and the various pools of nutrient elements are then better exploited.

Thus, with a high-diversity agroforestry system, a relatively closed nutrient cycle can develop and hopefully provide for sustainable productivity.

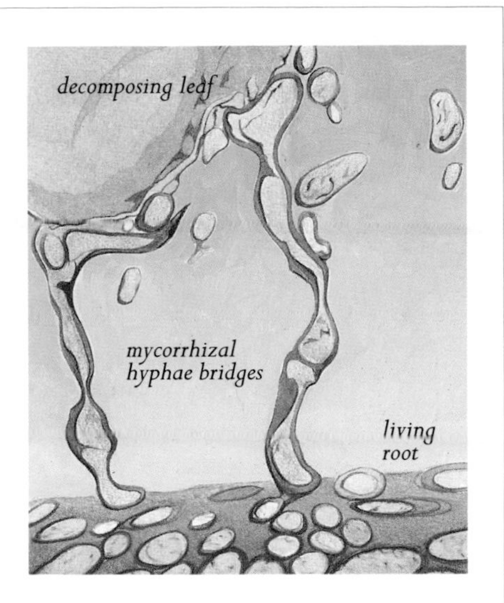

Close examination of the attachment between root tips and leaves has revealed "bridges" of fungal filaments, known as mycorrhizal hyphae, which facilitate the absorption of phosphorus and other minerals by the roots. The diagram above is a reproduction of an electron micrograph, showing the hyphae connection between living root tissue and a decomposing leaf.

Michael Yamashita

PART TWO

Forests and People

The indigenous people of the world's tropical rainforests face a fight for survival as their homes come under increasing pressure from encroachment and logging.

5 People of the Asian Forests

PAUL SPENCER WACHTEL

Asia is the most populous region on Earth, and one of the most powerful. It has been strategically important for centuries, and it has been an important source of natural products. But the region's real strength is the spirit of the people. Asians, while not necessarily paragons of virtue, have maintained deeply held and ancient beliefs about people's relationship with nature, which have lost ground only recently because of modern approaches to life that divorce people from wildlife.

Opposite. *Body decoration is a common feature among rainforest peoples everywhere, but perhaps nowhere has the art-form reached such a pinnacle of development as among New Guinea tribesmen. Elaborate head-dresses of hair, fur, feathers, leaves, and other materials are set off with bold and colorful painted facial designs. This highlander also wears a* bi pane *(an ornament made of shell and threaded through a hole cut in the septum between the nostrils) plugged with beeswax to hold it in place.*

A Delicate Balance

Think of some of the problems of modern Western life: pollution, traffic, crime in the streets, drugs. Then consider, for a moment, the very different problems faced by Mrs Badu Hamid of Gunung Labu, Sumatra. Her husband was eaten by a tiger.

The man's fellow villagers demanded retribution, but knew that tigers were one of Asia's threatened species and it is illegal to kill them in Indonesia. And, there are hundreds of tigers out in the forest behind the farms—so how could they be sure they got the right one? The people of Gunung Labu asked the Indonesian Nature Conservation Department for help, and the government officials called in an expert: Pak Suparno, a *pawang harimau*, or tiger magician, who is on a retainer with the government agency. Pak Suparno built a cage in the forest, baited it with a village dog, sang *pantuns* to the forest spirits, burned incense, and walked into the village the following morning to announce that he had caught the offending tiger. The animal was shipped to the Bengkulu Zoo, where it spent the rest of its life. When it died it was stuffed, and it now snarls harmlessly inside a glass cabinet in the Kerinci-Seblat National Park headquarters in Sungai Penuh.

Forest dwellers are expert practical botanists. This Penan hunter in Borneo has carefully blended the toxic properties of a number of plant species to produce the poison with which his blow-gun darts are tipped. With such a weapon he can bring down an animal as large as a gibbon at ranges of up to 50 meters (160 feet) or more.

Leimbach/Robert Harding Picture Library

Of course, not all people of the Asian forests are menaced by tigers, but the tiger, like some humans, has been pushed into the far corners of Asia, and both face an uncertain fate. Who are the threatened people of the Asian forest? They are the few thousand true jungle nomads who wander the primary rainforests of Malaysia and Indonesia. They are the vastly more populous hill tribes, who practice slash-and-burn agriculture at the forest's edge. And they are the millions upon millions of rural farmers who may not go into the forest very often but whose lives are influenced by forest animals and spirits, who rely on water that originates in the forested hills, who hunt animals that live in the forest, whose crops are eaten by those same beasts, and who obtain medicines from wild plants.

Historic Migrations

While Europe languished in the Dark Ages, great cities and sophisticated empires flourished in the rich rice-growing river valleys of China, India, Indochina, Thailand, Burma, Sumatra, and Java. At the same time, the remote hills and islands supported hundreds of tribal cultures based on shifting cultivation or hunting and gathering, each with its own customs and language, and each affected to various degrees by the rice-growing civilizations in the river valleys. This resulted in a rich mixture of major religions—Animism, Buddhism, Christianity, Confucianism, Hinduism, and Islam. All are still important forces in the region.

A good place to start is the east coast of Peninsular Malaysia, where leatherback turtles now share a stretch of beach with Club Med holiday-makers. That beach was high and dry as recently as 10,000 years ago. In fact much of the South China Sea didn't exist in those days, and it was possible to walk from Kuantan to Kuching, Bangkok to Brunei, Singapore to Surabaya. During the most recent ice age, at its coldest just 18,000 years ago, the low temperatures locked up so much water in the polar regions that the tropical sea level was lowered by some 120 meters (400 feet), and there was dry land where today the shallow South China Sea covers colorful coral gardens.

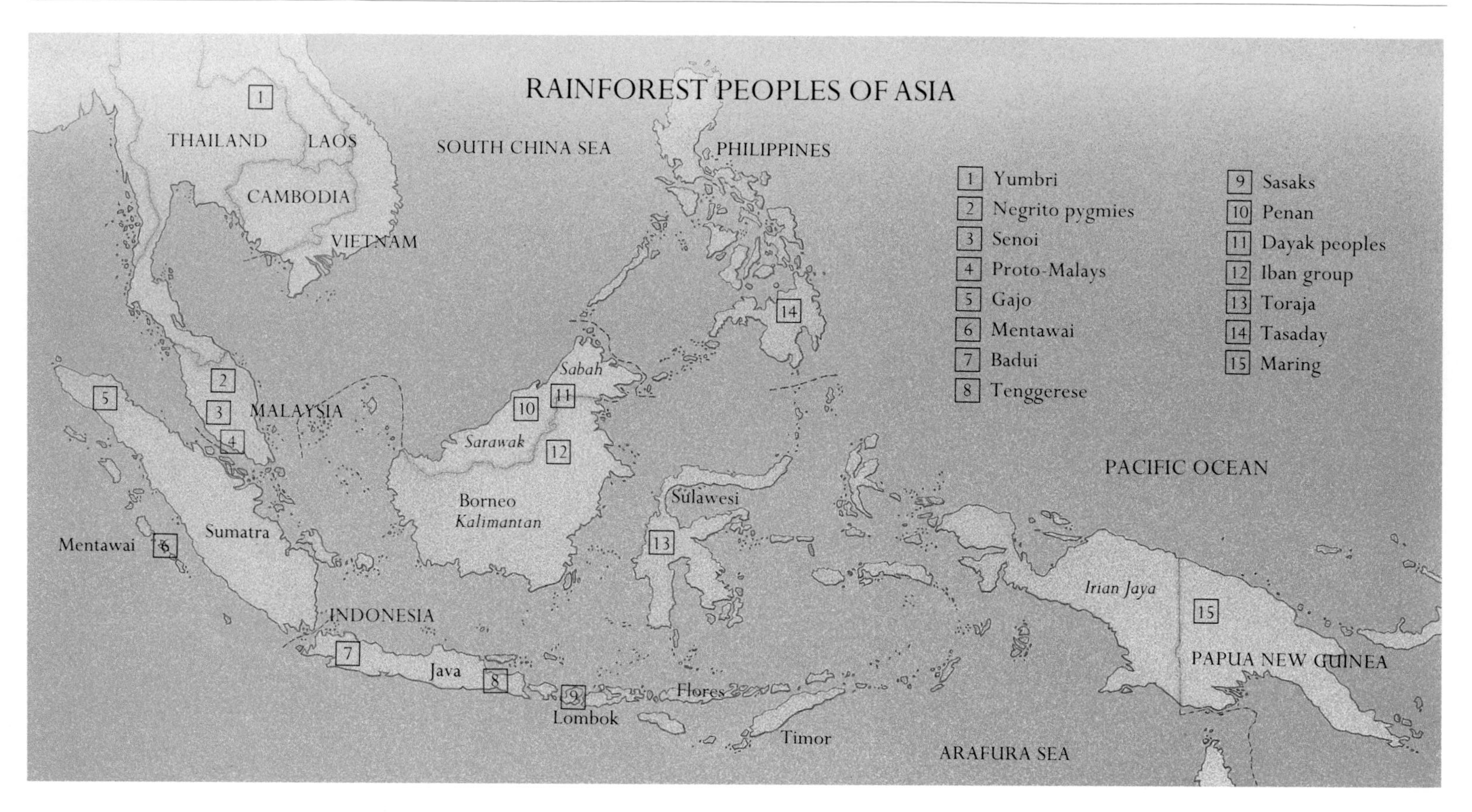

This map shows the approximate location of some of the peoples inhabiting the rainforests of Asia.

Opposite. *Armed with bows and arrows, a trio of hunters sets off on a bird hunt in the southern highlands of Papua New Guinea. Large mammals are relatively uncommon in New Guinea forests, and in their absence birds provide much of the people's protein. The feathers are also highly prized and are used in elaborate head-dresses and other body ornaments.*

But the low sea level was just part of a long cycle of shifting coasts. When the great polar ice caps melted, as they have done about 20 times over the past two million years, the rising sea level flooded most of the land, isolating the large islands of Borneo, Sumatra, and Java, along with thousands of smaller ones. Today the sea level is about as high as it has ever been, and Southeast Asia has only half as much land as it had when the poles were at their iciest.

Competing Lifestyles

The cycle of changing sea levels had important effects on the people living in the region. Humans, unlike many of the animals and plants, were able to cross the seas. Nevertheless people did not disperse evenly. Judging from fossil remains of *Homo erectus*, humans must have arrived in Java at least one million years ago, and it appears that they evolved into modern *Homo sapiens* while the sea levels changed around them. The early people were hunters and gatherers, like their African cousins, and the hunting was best when the sea levels were lowest. Most scientists agree that during those times the climate was cooler and drier, and most of the exposed land was covered in savannah grassland and monsoon forest.

Two changes occurred when the sea level rose. Islands were formed (and the length of coastline increased) and the area of tropical rainforest expanded. This challenged the ability of hunter-gatherers because many of the best sources of meat died out. Instead of being able to hunt relatively abundant large mammals on the savannahs, men were forced to hunt monkeys, mouse deer, and birds in the forests, a more time-consuming task. Scientists speculate that women also reduced their hunting, instead devoting more time to collecting a wide range of plants and animals.

An increased reliance on plants probably led to the first stages of agriculture about 10,000 years ago. One theory suggests that agriculture originated in Southeast Asia rather than the Middle East. The evidence for this theory came from a revolutionary discovery made by Dr. Boonsong Legakul and Chester Gorman in what is known as Thailand's Spirit Cave. The cave was used between about 12,000 and 6000 BC, and through careful detective work, Legakul and Gorman found that its inhabitants had a varied menu which included several varieties of beans, cucumbers, water chestnuts, almonds, peppers and bottle gourds, accompanied by fish and meat such as turtles, bats, monkeys, squirrels, rats, deer, and pigs.

It is clear that the people of Spirit Cave exploited a wide variety of plants, many of which are still found in the monsoon forests nearby. More importantly, these fruits and vegetables thrive in small clearings in the forest, an observation that led Gorman and his teacher, Wilhelm Solheim, to argue that these openings may have been small clearings or gardens made by the people of Spirit Cave. This would make them among the first people in the world to experiment with farming.

Pushed into the Forests

The history of Asian rainforest people is one in which successive waves of people took over prime lands and pushed the generally less sophisticated inhabitants further and further inland.

The original inhabitants of Southeast Asia were dark-skinned, frizzy-haired, broad-nosed Australoids, some of whom spread from southeastern Asia into Australia some 40,000 years ago during a low period in the sea-level cycle. They certainly used plants extensively, but they were not farmers. They had a complex array of stone tools, but no pottery or woven cloth. The Australoids were pushed into isolated corners of the region because they were unable to compete with later immigrants, who arrived with cultural and ecological innovations that enabled them to support much higher populations on the same amount of land.

The numbers of hunters and gatherers declined but they survive today as the Orang Asli (or "original people") of Malaysia and southern Thailand, the Andaman islands, the Philippines, eastern Indonesia and New Guinea (where they subsequently learned to cultivate yams and sweet potatoes). These descendants of the real natives of the region are treated with disdain and worse by the more sophisticated hill tribes and the even more sophisticated people who conquered the valleys and coasts.

The earliest farmers—the people who first made life miserable for the Australoids by pushing them into the deepest recesses of the forests, where they remain today—arrived relatively recently, coming into Thailand and Indochina only within the past 7,000 years or so, and moving into Indonesia even later. Called "Proto-Malays", they came overland from northwest India or Burma and had rich brown skin, a long head with wavy hair, and rather Caucasoid facial features. They pioneered the domestication of plants, and it is likely, but not proven, that the inhabitants of Spirit Cave were early groups of Proto-Malays. The fact that there is a discrepancy in dates between the Spirit Cave evidence (10,000 years ago) and the estimated arrival of the Proto-Malays (roughly 7,000 years ago) indicates the difficulty anthropologists have in making firm statements about prehistory.

Some Proto-Malays, among them the Yumbri of Thailand, the Tasaday of the Philippines, and the Penan of Borneo, lost the skills of agriculture and reverted to hunting and gathering. Others, such as the people of Mentawai island off the western coast of Sumatra, supplemented their hunting and gathering with various forms of simple agriculture. The best known of the Proto-Malays—the Gajo of Sumatra, the Tenggerese of Java's eastern uplands, the Sasaks of Lombok, the

Michael Yamashita

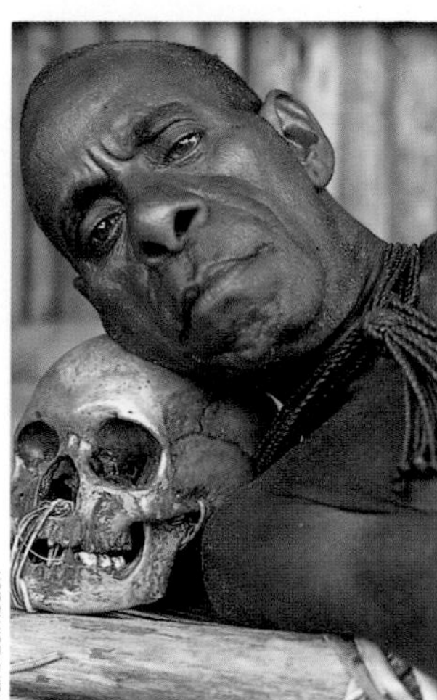
Claire Leimbach

An Asmat tribesman from Irian Jaya, New Guinea, rests on the skull of an ancestor, which he believes will protect his wandering spirit while he is asleep and ensure its safe return to his body when he wakes.

Opposite. *The gathering of the clans: Chimbu highlanders at a pig-killing ceremony in Papua New Guinea. Requiring little in the way of sophisticated husbandry techniques and thriving on the same diet as humans, pigs hold a special place in the lives of the native peoples of Papua New Guinea, who have few other convenient sources of protein. Pigs are also prized as status symbols and indications of wealth, and a pig feast is cause for great celebration, attracting tribes from near and far.*

Dayaks of Kalimantan and Sarawak, and the Toraja of Sulawesi—have mixed economies dominated by shifting cultivation. Invading shifting cultivators can move quickly; in one well-known case from recent times, an Iban group in central Borneo moved 300 kilometers (190 miles) in a single lifetime. To some extent these people can be considered "forest people", like the Orang Asli, as their lives are inextricably linked to the fate of the rainforests.

The most successful people in Asia have been the most recent arrivals, the "Deutero-Malays". They left southern China and traveled mainly by sea down the Indochinese coast to Malaysia, and to Indonesia via the Philippines. They have yellow-brown skin, a broad head and flat face, straight black hair and distinctly Mongoloid features and are today the dominant people of Southeast Asia. During the relatively short period from 3000 to 1000 BC, these agriculturalists used their knowledge of rice cultivation, pottery, and improved stone tools to spread throughout the region. In the process they split into numerous local groups and adapted to local conditions. The Proto-Malays and Deutero-Malays (often lumped together by anthropologists and called "Austronesians") also took to sea in outrigger sailing canoes, eventually settling new lands as far away as New Zealand, Easter Island, Hawaii, and Madagascar.

The Austronesian farmers were efficient immigrants. They expanded into fertile new environments that were occupied by sparse populations of Australoid hunters and gatherers. Population growth rates at this time could well have exceeded 3 percent per year, which is comparable to the rapid increase in population during the past few decades.

Common Belief Systems

Most urban dwellers believe that people are people and animals are animals. But the people of the Asian forests believe that a close spiritual tie binds wildlife with humans. All creatures have ethereal and eternal souls that can, and must, take refuge in a body. When one body dies, the soul can easily go to another body, and people and animals can exchange souls. This interchange may occur while the human is alive and be as temporary as a bad dream, or it might happen after the human body dies and the soul chooses an animal body as its new residence. As a further expression of this belief, most Austronesian cultures practiced headhunting because it was a means of capturing "soul energy", which could be transferred from one body to another. This "soul energy" was also harnessed as a hunting tool and symbolically to transform people into animals.

Belief in soul transfer (technically known as "metempsychosis") affects the behavior of people toward forest

Danny Lehman

animals. Tigers, as dominant, fearsome, nocturnal carnivores, are the animals most commonly associated with soul transfer. In areas where man-eating is a serious problem, people believe that the soul of a person eaten by a tiger becomes the soul of the tiger. Also, most groups have supported inspirational priests, or shamans, who claim the ability to communicate with spirits through trances. Shamans often claim tigers as their closest associates; during a trance the shaman becomes

the tiger-spirit, or were-tiger, by casting off his rational human demeanor to enter the world of magic and receive the revelations of the spirits. It is clearly in the religious leaders' interest to promote strong respect for such spirits and, of course, their own ability to deal with them.

Virtually all Austronesian societies have divided their deities into male sky gods (often represented by birds) and female earth gods (usually in the form of snakes).

PLANTS AND PEOPLE

In Sarawak, Borneo, a forest-dwelling Penan taps a tree to extract milk-sap. He collects it in a gourd and mixes it with other toxic substances. Later that day he dips his blowpipe darts into the poison, and sets forth to hunt. Meanwhile his wife gathers leaves to brew into a tea to treat her son's fever—the plants are her pharmacy.

Around the world more than 6,000 plants are regularly used in traditonal, folk, and herbal medicine.

MESSENGERS OF THE GODS

PAUL SPENCER WACHTEL

THE SEVEN EMISSARIES OF SINGALANG BURONG

Tuai Rumah Renang, the headman of Keranan Pinggai longhouse in Sarawak described the importance of omens: "When we are sent omens, it means that we are noticed by the gods. This is a thing of great importance. It is only by being noticed that we are able to excel at whatever we turn our hand to. If the gods 'see' and take an interest in us—then we are likely to become wealthy, reap large harvests year after year, and escape sickness. If we are not noticed by the gods, then we are 'without luck'." The gods show their interest by sending omens to provide guidance, and people must reciprocate by respecting the omens and showing gratitude.

Many of the bird beliefs in modern Borneo resemble those of classical Rome. Just as the Romans considered birds the messengers of Jupiter, so the tribes of Borneo consider birds the intermediaries between themselves and Singalang Burong, the Iban sun-bird, their god of war and most powerful deity. As befits the chief, Singalang Burong has seven emissaries to handle the day-to-day affairs of terrestrial humans. Like many Asians, he tries to keep things in the family: six ambassadors are his sons-in-law and one is an unmarried dependant. None of the birds are large or particularly colorful, but together they reflect the most important facets of human endeavor.

According to the World Health Organization, traditional medicine provides the primary health care for 80 percent of people in the developing world. As Harvard professor Richard Evans Schultes (the "father of ethnobotany") puts it: "Through most of man's history, botany and medicine were, for all practical purposes, synonymous fields of knowledge." Skeptics may scoff that many traditional medicines are worthless. Yet it is a fact that 74 percent of the plant-derived chemicals now prescribed in the West were used by other societies for similar purposes long before their discovery by science.

It is easy to appreciate the value of plants to the people of the rainforest. Yet in Southeast Asia, as elsewhere in the tropics, deforestation is taking place at a dramatic rate. The result is that rainforest people are losing their lands. In many cases they have no choice but to assimilate into the mainstream. Some groups, such as the Penan, are fighting. Other groups go quietly. To be fair, many people of the rainforest welcome the move towards a more stable and sophisticated life. They want the dream of motorcycles and jobs, education and good clothes, sewing machines and televisions. The point is that no one should impose a lifestyle change.

But for those forest dwellers who want to make their own decisions and who wish to be left alone, the future looks bleak. Little by little, the forests are being eaten away by industry. The human tragedy is serious and of concern to many, and the botanical tragedy is becoming equally important. The two are inextricably linked.

Forest dwellers are the world's most knowledgeable economic botanists. The knowledge of which plants can cure specific diseases is often unwritten and is passed from parent to child—that is, if the child is interested in learning the old people's ways. Scientists have tried to tap this knowledge, but so far only about 10,000 of the 250,000 species of flowering plants thought to exist have been examined by modern investigators. This statistic is critical when you consider that the World Wide Fund for Nature (WWF) estimates that if current destruction rates continue, about 60,000 plants (one in four) will be threatened or extinct within 60 years.

A Troubled Future

Today, some isolated peoples are being encouraged to join Deutero-Malay society, generally on terms set forth by dominant groups. The Malaysian government provides schools, roads, and a flying doctor service for the Orang Asli of Peninsular Malaysia in an attempt to bring them into the mainstream of Malaysian society. In Java, Indonesia, the government has had moderate success in convincing about 60 families of the Badui tribe to move into new villages. The Badui are a determinedly isolationist group with a reputation (like most isolated groups) of having powerful magic. They are so strong-willed that Indonesia's President Suharto is said to have wanted to meet their headman to obtain a charm. The President was willing to travel to the outer limits of the Badui lands, but the Badui elders said that if he wanted to meet them he must walk a day through the forest into their inner sanctum. They never met. The Badui have managed to maintain their independence although they live relatively close to Deutero-Malay population centers. Nevertheless, the Indonesian government has convinced some Badui to leave the compounds they have occupied for some 500 years and learn Indonesian and watch television like most other people of this vast country. As the Minister of Social Affairs explained, the schools, clinics, markets, and other amenities of a civilized life would provide the Badui with "the benefits of being Indonesian". This is probably a fair summary, but it should also be noted that by joining the mainstream the Badui will likely lose

Anthony Howarth/Susan Griggs Agency

An Iban elder. The Iban tribe of Borneo are members of the Dayak group of peoples and, like other Dayaks, they live in longhouses—motel-like structures on stilts that are home to entire villages.

the benefits of being Badui. Few isolated groups have been able to retain very much in the way of traditional lifestyle once they leave the security and mutual-support system of their homelands.

TRANSMIGRATION PROGRAMS

Throughout Southeast Asia, the Mongoloid population grows at the expense of the Australoid. The most significant example of this cultural and geographical encroachment is the Indonesian transmigration program. Indonesia is the world's fourth most populous country (after China, India, and the United States). More than 100 million of the country's 170 million people live on Java, an island with just 7 percent of the nation's land. The transmigration program—claimed to be the largest voluntary, government-supported resettlement program in the world—aimed to move millions of Deutero-Malays from Java, Madura, Bali, and Lombok to the more sparsely populated areas of Sumatra, Sulawesi, Kalimantan, and Irian Jaya. The objective was to reduce population pressure on Java and open up frontiers for development, partly by giving the new settlers land.

The results are mixed. Although some two million people have moved, most during the 1980s, this hardly makes a dent in Java's population. The objective of "developing" the outer islands has been more successful, since transmigration is accompanied by roads, communications, electrification, schools, markets, and other amenities. Another benefit, according to World Bank studies, is that transmigration has created at least 500,000 jobs, bringing many people into the labor force for the first time. Also, it seems that the lives of most migrants have improved after moving. Not everyone agrees that this is good "development". Some critics argue that the transmigration program amounts

John Launois/Black Star

Victor Englebert/Susan Griggs Agency

Rice is the staple crop of the Dayak peoples of Borneo. Families work the plots together, the men preparing the ground, the women sowing. Astronomy determines the planting cycle: the land is cleared when the Pleiades rise above the horizon in June, and the crop planted in September when the Pleiades reach their zenith. The plots become unusable after three or four seasons and new ground must be cleared.

to little more than Javanese imperialism. The people who suffer most from the program are those who have the least power to begin with—the forest dwellers and other people who are relatively low in the hierarchy.

Transmigration has led to environment problems as well. Too often the transmigration sites have been located on poor land. Sometimes watersheds, which provide essential water, have been cut down to clear the land for settlers. (This was one reason the World Bank contracted the World Wide Fund for Nature to help the Indonesian government manage a forested watershed near a transmigration site in Sulawesi.)

Sometimes the government appropriates forest land that the original occupants say belongs to them, leading to debates that the forest dwellers inevitably lose. And sometimes the program results in quite unexpected circumstances. For example, the central government of Jakarta established a settlement in a sparsely populated area in southern Sumatra, called Air Sugihan. Unbeknown to all concerned the village was placed in the middle of an ancient elephant migration route. The new inhabitants found that not only did they have to contend with the trials and tribulations of a first crop, but they also had to keep the elephants out of their fields. One night, the inevitable occurred—the elephants stampeded the village destroying anything and everything in their path. Days of reconstruction were required to return the village to its former state, and there was no guarantee that the elephants would respond to a new government effort to move them to another stomping ground. One section of the Indonesian government was responsible for the transmigration program, another for conserving elephants. One can only imagine that the original people of the Sumatran forest would have been wise enough not to build their homes on an ancient elephant highway. ■

In 1971, the now-famous Tasaday were found living a stone-age hunter-gatherer existence in Mindanao, the Philippines, just 25 kilometers (15 miles) from the nearest road. There has been much controversy about whether the Tasaday are "genuine", but many of the world's anthropologists were quite prepared to accept that an unknown tribe had been discovered.

6 People of the African Forests

JUSTIN KENRICK

For centuries Europeans have thought of the tropical forests of Central Africa as being, in the title of Joseph Conrad's story, the "Heart of Darkness". The reality is very different. In the heart of the forest—in parts of Zaïre, Cameroon, Gabon, Congo, and the Central African Republic—live some of the few remaining hunter-gatherer peoples, to whom the forest is a secure and bountiful home. By contrast, the farmers who long ago settled along the waterways and tracks that cut through the forest see it in many ways as a hostile environment. They cut the forest so they can grow crops and survive; and then when the soil is exhausted they move on to clear and plant new areas in the forest, allowing secondary forest to reclaim their old fields.

Opposite. *A barbecue, Mbuti-style. The Mbuti of the Ituri Forest in northern Zaïre are partly nomadic and have little use for possessions. They live in groups in simple huts formed of woven, pliable saplings and thatched with leaves. Gregarious and mild-mannered, they lack chiefs, and all decisions affecting the band are made democratically.*

Economic and Cultural Exchange

Until 50,000 years ago the Central African moist forests covered one-and-a-third times their present area. They gradually contracted, mainly due to the hunter-gatherers' growing use of fire and also in response to global climate changes, until the present level of aridity was reached about 3,000 years ago. During this period, people on the southern border of the Sahara began to grow cereals, while yams and palms began to be cultivated in the forests of West Africa, where the relatively fertile soil eventually led to thriving agricultural economies such as that of the Asante (Ashanti).

The meeting place of these two agricultural traditions occurred in what is now Cameroon; it was here that farmers began to experience a shortage of land. The existing system of food production meant that villages needed to move to new sites once or twice every decade, the best new sites being the borders of natural clearings and rivers. More than 3,000 years ago these farmers spread into the Central African forest: east through the northern woodlands of the Ubangi basin, and south through the forests to the Zaïre basin, until they reached all of Central and southern Africa.

Gradually these farming peoples were able to establish highly organized cultures and trade networks through much of the forest. They also began exchanging with forest peoples—agricultural produce, iron, and pottery artifacts for forest produce. The introduction of bananas and other plants that were easy to grow enabled farmers to settle in new areas and more easily produce food surpluses to exchange with the hunter-gatherers.

The Mbuti pygmies consist of both Efe archers and Mbuti net hunters. Here, an Efe man adjusts bundles of arrows. These arrows, or manzi, *as the Efe call them, are fashioned from palm fronds then placed over hot coals to harden the tips and dry poison into them.*

Robert C. Bailey

The Forest People

The forest people believe themselves to have been the first inhabitants of the forest, a belief shared by neighboring farmers. The Bantu farmers, for example, tell stories in which their ancestors arrived from somewhere else, and the pygmies, as the first in the land, were the knowing guides who taught them how to cope with their new environment. Today the farmers express an ambivalent attitude towards the people of the forest, and the relations between them range from virtual serfdom to freely-entered-into exchange relations.

The Baka of southern Cameroon and the Aka, living in the Central African Republic and northern Congo, are separated by the Sangha River. Thousands of kilometers to the east are the Mbuti of the Ituri Forest in northern Zaïre; they consist of both Efe archers and Mbuti net hunters. The combined population of these separate groups is probably only about 100,000, and population density in the forest is not high—the Mbuti average about one person for every 4 square kilometers

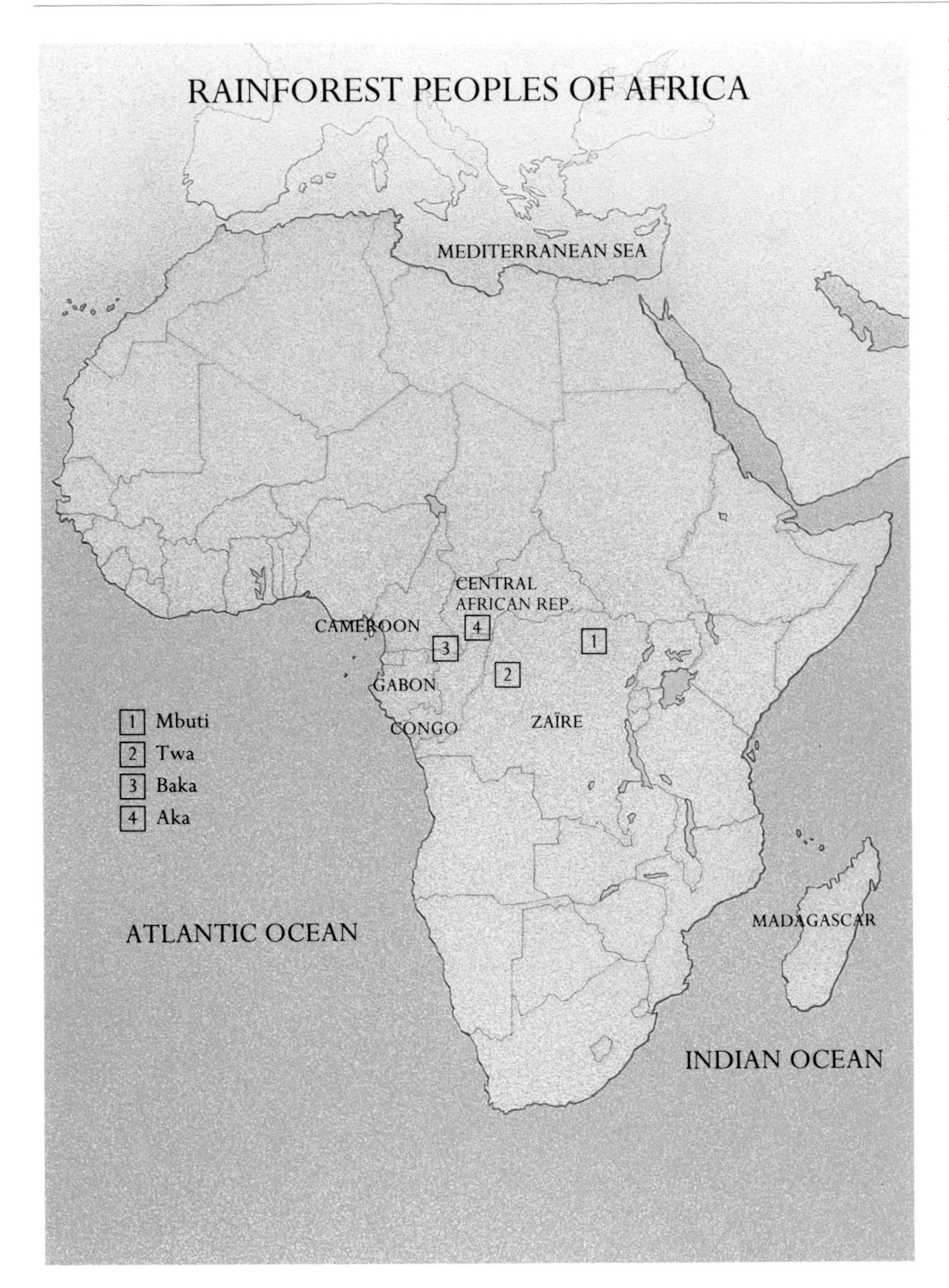

This map shows the approximate location of some of the peoples inhabiting the rainforests of Africa.

(1½ square miles). There are approximately 70,000 more forest people, including the Twa, in the central basin of the Zaïre River.

To varying degrees the Baka, the Aka, and the Mbuti continue to exist as hunter-gatherers. Although they are separated by distance and language, they share a strong cultural identity based on their relationship with the forest. They perceive their environment to be plentiful and benevolent, and recent studies have shown them to be nutritionally better off than most other peoples of sub-Saharan Africa. The languages they use are largely derived from those of neighboring farmers. However, the existence of a shared core of words, particularly plant names, which are not borrowed from neighbors, suggests that they once had a common language.

The forest peoples are small in stature, and this is one way in which they are superbly adapted to their environment. Research has shown that African forest people are smaller at birth than their neighbors and grow at slower rates throughout childhood. For example, an Efe child at age five is the height of a two-and-a-half year old American girl. Forest people are not only smaller than their neighbors, they also have relatively less muscular mass. The same is true of forest mammals in comparison with their relatives in the savannah, and it has been suggested that this small stature is an evolutionary adaptation to their environment. Certainly, their size helps them to take cover in dense thickets and to follow the tunnel-like trails made by the duikers and antelopes they hunt. Their light muscular frame is also suited to tree-climbing.

Traditional Ways of Life

In general, the forest people spend long periods in the heart of the forest. They live in bands of between 15 and 60 people, hunting for meat, gathering plant foods, and collecting honey. Everything they own has to be carried when they move to a new hunting camp, so there is considerable advantage in having few possessions. What they do have in abundance is an intimate knowledge of the forest: the ability to read animal tracks, to know the flowering and fruiting cycles of forest plants, to locate a bees' nest from the flight of a bird. Forest people know the individual properties of thousands of plants and make use of them to eat, to make poisons, to dull pain, heal wounds, and cure fever.

To set up camp in the forest they clear a small area, leaving large trees in place. The women weave small round huts from pliable saplings, which they then thatch with leaves.

When the group moves on to gather and hunt in a new area, the camp is soon reclaimed by the forest. Invariably the forest people spend much of the year near a village, where they may exchange meat, mushrooms, or honey for manioc and palm wine. They also work in the villagers' gardens, apparently as subordinates, although they are usually free to return to the forest at any time. The Baka tend to move into the forest at the start of the rainy season when the wild mango trees start to fruit, the harvesting of the villagers' coffee and cocoa plantations is over, and game in the vicinity has become scarce. For three to four months of the year Baka groups move through the forest gathering mushrooms, yams and fruits, and hunting animals. The Aka may walk for two days, 14 hours a day, before setting up their first hunting camp, and they can spend up to eight months deep in the forest, hunting with nets. In contrast, the Efe archers of the Ituri Forest never go more than eight hours from their Lese farmers' villages.

The exchange between foragers and farmers allows the farmers to acquire meat to improve their diet. Generally, an Aka or Mbuti family will always trade with the same village family. And since there are many more villagers who want meat than there are Mbuti or Aka to provide it, the forest people are able to pick trading partners who will treat them well.

Some anthropologists argue that the forest people are able to maintain a reliable supply of starchy foods only by exchanging the meat they have hunted for crops grown by farmers. They therefore suggest that the forest people could never have existed independently in the forest but must have originally lived in areas bordering the savannah, moving deeper into the forest only when they could settle alongside the Bantu farmers. But during the Simba Rebellion in Zaïre in the 1960s the forest people retreated into the forest and not only survived for years while the civil war was raging around them, but also hid and fed many village farmers.

Probably the most useful item the forest people obtain from villagers is iron for axes and for the tips of spears and arrows. Even these items are not essential, because they also use fire-hardened spears, arrow tips impregnated with poison, and bamboo cutting utensils. Likewise, they can create cloth in the old manner, by hammering the bark of certain trees into suppleness.

In economic terms the trade relationship seems to be a choice, because both the foragers and the villagers are able to live off the food and goods of their own environment. Perhaps the underlying reason for continual trade is that it creates a relationship between the forest people and groups that might otherwise threaten their way of life in the forest.

The persistence of the forest people's way of life, despite more than 2,000 years of contact with farmers, demonstrates not only the attractiveness of the lifestyle but also the ability of forest people to relate to other cultures without letting that contact threaten their own cultural identity. The existence of the forest as their refuge is clearly crucial, but equally important are the advantages they see hunting and gathering have over farming. There is less work—they spend a much smaller proportion of their waking time collecting food than the farmers spend tending their crops—and because there is no worry about whether the harvest will fail, there is less uncertainty. If game is sparse in their territory they can easily move through the forest to another area.

Each band needs to cooperate in the challenging and entertaining activity of the hunt, and in childcare and sharing food. The size of a band constantly grows or shrinks in response to the availability of food, with individuals and families moving between bands.
In the forest there is clean water, and the high canopy protects them from direct sunlight. Mosquitos, which live in the canopy, virtually never come below it, except beside rivers. By contrast, unclean water in the villages spreads dysentery, and the lack of forest cover encourages mosquitos, so there is more malaria. Because cultivators rely on protecting and owning their land and produce, questions of ownership and inheritance are a cause of conflict. For the hunter-gatherer these are not problems: all have equal access to resources, because they need each other in order to harvest the forest successfully. There are no chiefs or formal hierarchies; they are fiercely egalitarian, and any important disputes can ultimately be resolved by one party simply moving to join another camp.

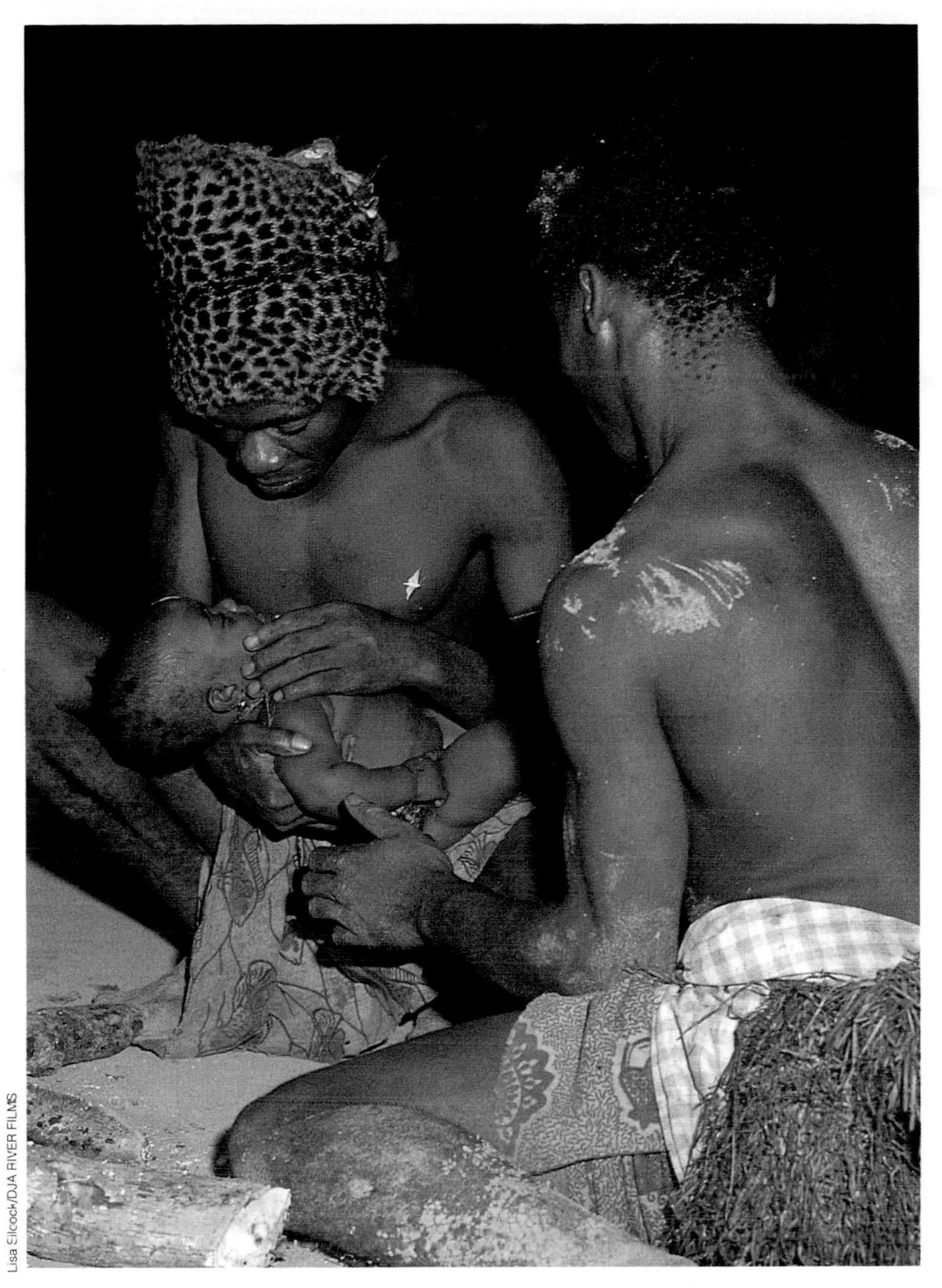

Lisa Silcock/DJA RIVER FILMS

A group of Baka pygmies care for a sick child. Despite a common portrayal otherwise, forest peoples are not helpless in the face of illness and disease. They know the medicinal properties of thousands of rainforest plants, many of which form the foundation of Western drugs, and can deal successfully with a wide range of ailments.

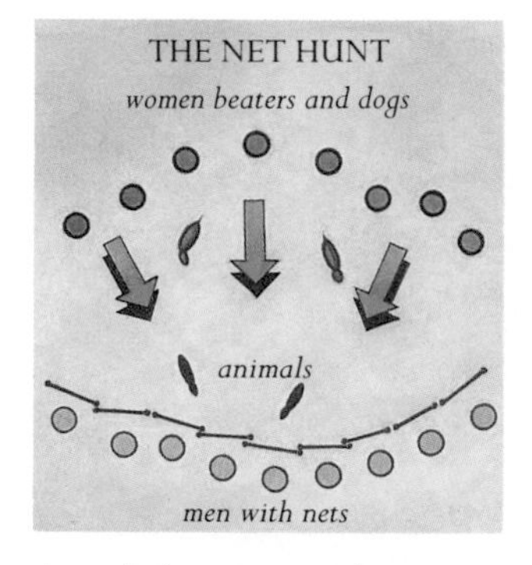

A simple diagrammatic depiction of the Mbuti net hunt. Each hunter sets up his net beside another so that together the nets form a semi-circular trap, which may extend up to a kilometer. The women, helped by dogs, noisily beat the bushes and drive any animals toward the nets. The men stand poised to spear the animals once they are entangled.

Gathering and Hunting

Among both the Aka and the Mbuti, gathering occurs primarily during the course of the net hunt. Women do most of the gathering and use baskets on their backs to carry the various forest products. The Aka, for example, may collect roots, nuts, fruit, leaves, mushrooms, honey, caterpillars, vines of *kusa* for string, *ndemele* to make poison for arrows, and *bongbou* to weave into baskets. Whatever is gathered is taken back to the individual's family, although some sharing between families occurs. The greatest volume of food comes from forest roots, and from village manioc acquired through trade, but the most highly prized food is honey. Sometimes villagers venture into the forest with the Aka to hunt and gather, and although they are not at home there, they are not excluded from the forest in the way that the Mbuti have, until recently, managed to exclude "their" villagers.

Hunting involves both men and women, although the men generally take the lead. Different peoples employ different hunting methods. The Efe in the Ituri Forest spend most of the year in small bands using bows and arrows, and come together in larger groups for the honey season. Among the Baka and Aka the crossbow has gradually replaced the bow and arrow, its use having spread from West Africa and become part of the Aka culture in the past 30 to 40 years. Like the bow and arrow the crossbow is used primarily to hunt monkeys, usually in the morning when monkeys are more active in their search for food. If the arrows have been treated with poison, it is one that paralyses the monkey but does not harm the meat.

For both the Mbuti and the Aka, the net hunt is a basic hunting technique. Both peoples hunt daily except during rain or heavy mist, when the nets become moist and liable to tear. The nets are about 45 meters (150 feet) long and about 1 meter (3 feet) wide and are made from a fibrous material found in the bark of *kusa*—a forest vine of considerable strength. The net hunt depends upon cooperation. It is not feasible with fewer than seven nets, and becomes unwieldy with more than 25 because their total length is then more than a kilometer. This sets the upper and lower size limits of the hunting band because every married man sets up a net for the hunt and is responsible for making, repairing, and carrying it. Both the Aka and Mbuti shout and yell to scare animals, such as duikers and antelopes, into the nets. The Mbuti arrange their nets in a semi-circle; the women line up opposite the nets and noisily try to drive the animals into them; the men lie concealed just beyond the nets, waiting to kill. In contrast, the Aka arrange their nets in a complete circle; the men try to drive the game into the nets from the center of the circle and the women lie concealed just inside the circle.

Patrick Bordes/EXPLORER-AUSCAPE

A recent study by Barry Hewlett, an American anthropologist who has worked with the Aka for many years, demonstrates that husbands and wives cooperate in subsistence activity for the greater part of the day. One consequence of this is that fathers can develop sustained affectionate relationships with infants; Aka fathers spend 47 percent of their day holding, or within

an arm's reach of, their infants; and while holding the infant, the father is more likely than the mother to hug and kiss the child.

Aka fathers are involved in childcare to a far greater degree than any other society studied by anthropologists, although their involvement is as much a consequence as a cause of the sexual equality in their society. Many other foraging societies, in which fathers are not as involved in childcare, also have a high degree of sexual equality, which suggests that the egalitarian ideology and other features of such societies are equally important in maintaining equality between individuals and between sexes.

Other features that serve to maintain a vigorous equality among the forest people of Central Africa are the celebrations, music and dance, which include everyone in the camp and require a high degree of cooperation. The section that follows looks at these,

A band of Mbuti, armed with nets and spears, prepare for a net hunt, which is generally a communal affair involving the entire village. Every married man has a net, which is rather like the net on a tennis court, but much longer.

Lisa Silcock/DJA RIVER FILMS

A young Mbuti man weaves a honey-basket. Honey from wild bees is much prized by the Mbuti, and during the flowering season the hunters seek little else. Most is consumed on the spot, but what is left is carefully gathered into specially woven baskets and carried back to the village.

Opposite. *A pygmy woman and child gather wood in the Cameroon forest. Among many pygmy tribes in tropical Africa women and children are typically the gatherers. Hunting, on the other hand, involves both men and women, although the men generally take the lead.*

drawing in part on the superb work of the anthropologist Colin Turnbull, whose book *The Forest People* is a vivid account of his time spent living with the Mbuti people.

Mbuti Celebrations

Mbuti children grow up with a strong sense of security. This may have its source in the sense of the forest as provider, but it is manifest in everyday life. The nuclear family is important, but all mothers in a camp are called "mother" by a small child, and all fathers "father". Likewise, all adults of parenting age care for the child as a mother or father. While the child is still in the womb its mother sings and talks to it not with baby talk but in all seriousness and humor. Infants are breastfed for several years. Interestingly, this fact alone does not explain how the Mbuti manage to keep their populations low. Although sex is enjoyed by young unmarried people, they appear not to have children until after they are married.

As they grow, Mbuti children play games through which they learn values and skills. By constructing a swing out of a vine they learn timing and balance; one of them leaps off the swing just as another leaps on. In another game they all climb a small sapling until their weight causes the top of it to curve right down to the ground. At that moment they leap off and any child who has not been sufficiently aware to jump at the right moment gets a shock as the sapling flies back into place. During the day, when the adults have left to gather and hunt, the grandparents may use an old hunting net to teach the children aspects of the cooperation and mutual awareness necessary for a successful hunt.

The Mbuti believe that they used to be immortal until the first time one of them killed an animal. They also believe that children, who do not take part in the hunt, are purer and closer to the forest than the adults. It is the children's job to light the morning hunting fire, which the adults pass through as they leave the camp to cleanse themselves of the disharmony they cause by hunting and killing. As youths, Mbuti take on the important role of bringing disharmony into the open, where it can be resolved through argument, through humorous ridicule, or through involving the whole camp in singing and dancing.

The Mbuti boys undergo the villagers' *nkumbi* initiation into manhood at some stage between the ages of nine and eleven. This helps to forge strong connections with the village boys, who may become future trading partners, and it satisfies the villagers' need to feel they have some control over the Mbuti they believe they "own". The Mbuti themselves do not make the transition into adulthood until they marry, build their own hut, and have the right to make and own a hunting net.

The potential for conflict between the sexes is first explicitly dealt with in the *elima* festival, when a girl celebrates her first menstruation. The *elima* includes a "tug of war" between men and women: when one side starts to win, someone moves to the other side, imitating and ridiculing the gender they have joined. This swapping of sides continues until everyone is imitating the opposite sex, and they all collapse in laughter. Nobody wins, but the conflict between the sexes is brought out into the open, and the value of cooperation, equality, and humor is asserted.

When there is a death or serious disharmony in the camp the elders call for the *molimo* ritual, which involves singing and dancing night after night. But it is up to the youths, who carry the wooden *molimo* trumpet into the camp, to decide whether the singing and spirit is strong or half-hearted. If it is strong then the sounds they make through the trumpet will be the soft growls of the leopard, the spirit of the forest. If it is not then they will rush into the camp sounding the shrill trumpeting of a stampeding elephant, tearing at huts, knocking everything flying, and reproaching the camp for not being in harmony with the forest.

Liquid Gold

For the Mbuti, the most highly prized treasure in the forest is honey, both the dark honey of stingless bees and the golden honey of stinging bees. The start of the honey season is marked by the honey dance, which is also an opportunity for different camps to come together to socialize, for young people to find possible marriage partners, and for individuals to leave one band and find another one for the next year's hunting. During the honey season, the Mbuti abandon the net hunt and break up into smaller groups.

Beehives can be 30 meters (100 feet) above the forest floor, and the Mbuti climb from saplings to larger trees, using vines and branches to get close to the nest. They carry a burning log wrapped in green *mongongo* leaves, which they blow as they approach the nest. The smoke stuns the bees, allowing the men to hack their way into the hive. What isn't consumed immediately is collected in a basket and carried back to the camp. It takes many days and many beehives to gather enough honey for trade with the villagers.

Colonialism's Legacy

Throughout the colonial period, forest peoples suffered intense pressure to abandon their forest life and become farmers. Central African States came into existence as independent nations as recently as 40 years ago, with boundaries artificially drawn up by the colonial powers, who largely ignored the cultural rights of marginal people. Sadly, the post-colonial era has seen few improvements in attitudes to these people. For example, unsuccessful attempts were made to "domesticate" the Mbuti of Zaïre in the 1970s. The forest peoples of Africa have little access to the democratic political processes that their counterparts in South America are beginning to enjoy. The government of Gabon even denies that there are any forest people in Gabon.

The Twa of western Zaïre were probably the most severely affected by the brutal policies of King Leopold of Belgium, who held the Congo (now Zaïre) as his personal fiefdom from 1885 to 1908, forcibly bringing the Twa out of the forest to settle by the roads. Initially they became suppliers of ivory to the colonial economy through their village neighbors. Later, the imposition of taxes and the use of force to ensure they worked as laborers producing cash crops left the Twa firmly under the control of agricultural communities, unable and eventually unwilling to return to living nomadically.

In the Congo the French attempted a much less successful "taming policy" with the Aka during the 1930s. Today in many parts of the Congo the Aka are said to be virtually slaves to villagers—although it may, in fact, be a more complex reciprocal relationship.

G.P. De Foy/EXPLORER-AUSCAPE

A pygmy couple poses for a portrait in their forest homeland of south-eastern Cameroon. A common feature of indigenous peoples everywhere is the selective adoption of Western dress. Superior durability and easy care of such clothes are as evident to the forest dweller as to the person from the West.

Many Aka move to Bantu villages for part of the dry season to help harvest the coffee plantations; others used to help villagers hunt elephants for ivory but have now taken up farming or work for timber concessions. The same is true of the Baka; the Cameroon government has had a program for the settlement and economic integration of forest people since 1980. In the Ituri Forest of Zaïre much of the pressure on the forest comes from the recent influx of people such as the Nande, dispossessed by the intense cultivation of cash crops and forced to move west and into the forest as shifted cultivators, coffee growers, traders, or gold prospectors. With the forest under intense pressure in all these areas, the bush-meat trade has increased dramatically. The Mbuti and Aka hunt for meat traders from the town, which means there is an overkill that forest animal populations cannot sustain. This would have been unthinkable a few years ago.

Where the forest has disappeared or been severely degraded, or the authorities have succeeded in turning the forest people into village farmers, those who continue foraging are often exploited by government functionaries. Whereas the traditional farmer–forager relationship contains a multiplicity of ties that are both economic and social, exploitation is based only on a single-stranded exchange, such as meat for money.

In fact the forest people already make an important contribution to the economy, culture, and health of their countries through their hunting, gathering, and

Robert Harding Picture Library

Robert C. Bailey

exchange; their music and dance; and their irreplaceable knowledge of the forest. If Central African governments could recognize this, then instead of becoming embarrassed about their "backward pygmies" they might see this cultural diversity as enriching rather than destabilizing. They might even begin to halt or reverse policies that exploit, marginalize, or actually destroy the lifestyle of the forest people. There is a chance of this happening as the region moves towards multi-party democracy and environmental awareness. Much depends though on the attitude of Western governments, which may no longer continue supporting unaccountable single-party dictatorships in the nations of Central Africa simply because they are anti-communist. There is a slim chance that human and minority rights issues, democracy, and environmental awareness will become more important in Western eyes than supporting governments offering cheap and abundant supplies of raw materials such as timber and minerals.

The Impact of Logging

During the past decade it has been Africa and not Southeast Asia or South America that has suffered the highest level of deforestation. A study by the Food and Agriculture Organization of the United Nations shows that between 1981 and 1990 Africa lost 17 percent of its rainforest, Asia 14 percent, and South America 9 percent. More than 75 percent of West Africa's forests have now been destroyed, and this has been

Mbuti villagers gather at an elima *festival, held when a girl celebrates her first menstruation. Such festivals include a variety of games, including one in which a tug of war is staged between the men and women of the village; nobody wins because participants continually shift to the losing side until utter confusion reigns and the whole tangle collapses in laughter.*

Robert C. Bailey

Moving camp in the Ituri Forest of Zaïre, a Mbuti woman carries the family possessions, a burden weighing as much as half her own body weight.

Opposite. *A successful Mbuti hunter returns home carrying his catch, a duiker. Hunters often operate alone, but more often in groups because the chances of returning empty-handed are reduced. The hunter first drawing blood gets the largest share, the rest of the kill being divided among the group according to strict convention.*

mainly because of European logging companies. Logging greatly increases migration into the forest, opening it up with new roads and destroying or severely degrading it. In the Central African Republic, 90 percent of the forest has already been allocated to European logging companies. Much the same is happening in the Congo, where the World Bank has been involved in financing logging. In Zaïre, logging by European companies is increasing dramatically, although the Ituri Forest is safe for the moment because it lacks the high-grade timbers favored by Europeans. Ninety percent of the large trees in the Ituri are *Mbau* (Gilbertiodendron), which is notorious for splitting when cut into planks; but even in the Ituri, sawmills are at work in the patches of mixed forest. Meanwhile in southern Cameroon, North American and European logging companies have built roads through large areas of previously untouched forest. Sedentary jobs and poorly paid work at logging camps mean less hunting time for the formerly nomadic Aka and Baka, and therefore often a lack of protein in their diet. Conflicts that would before have been solved simply by moving to another area are instead exacerbated. Sanitation becomes difficult, and the people become prey to parasites and malaria-carrying mosquitos. By settling permanently near roads and logging camps, the delicate reciprosities they shared with the villagers break down. Within the camp, individual wage earning, and inevitable financial selfishness, break down the egalitarian principles on which the culture was built.

The creation of logging settlements also increases the demand for meat and manioc. The middlemen bring in money, cannabis, and clothes to exchange for meat. These things are easier to transport into the forest than bulky but more essential village produce such as manioc. Tension occurs as outsiders disrupt life in the forest and diminish the traditional relationship of forest people with their neighboring farmers. The logging companies cut the forest to take high-grade timbers for Western markets, but they move on after a few years; their presence can dramatically influence the ability of forest people to adapt to stress in the future. Barry Hewlett describes how in a measles epidemic the Aka groups that had been most tied to logging camps had a much higher fatality rate among their children—the breakdown in their relationship with traditional farmers had led to a lack of nutrition.

There is a growing international consensus that unsustainable logging must be stopped, to save both the forest and the forest people. This could be achieved by reimbursing governments through "debt-for-nature" swaps (in which Third World debt is cancelled in exchange for the creation of protected-forest areas, or national parks). Another possible solution is to provide financial incentives for companies to pursue sustainable timber production in areas that have already been devastated. It is vital, however, to recognize that sustainable logging can only be achieved if it takes into account not just the need to stabilize logging (so that it can maintain a sustainable yield of timber into the future) but also the needs of the local people to establish a secure livelihood and maintain the vital functions of their local ecosystem. On these terms, sustainable logging is only possible when it is the local people themselves who benefit from the revenues of logging and other forest products, and from the maintenance of their long-term financial, social and ecological security. In this sense the protection of the environment goes hand in hand with democratic decision-making and the alleviation of poverty.

The Future

The future of the forest people of Africa is uncertain, particularly because the egalitarianism, autonomy, and flexibility that characterize the forest people's social relations also mean that it is difficult for them to unite against external pressures building up against them. As recently as 10 years ago, anthropologists working with them assumed that their way of life would disappear along with the forest as African countries exploited their natural resources to the full. Today, however, one could argue that their chances of cultural survival do not look much worse than our own. Throughout the world, and perhaps especially in the West, we are realizing that we cannot survive without a viable ecosystem and that central to the balance of the world's ecosystem is the existence of the tropical forests. ■

Robert C. Bailey

7 People of the Central and South American Forests

THEODORE MACDONALD, JR.

In a hunting trail in Peru's Manu National Park, a startled Machiguenga Indian meets a reclusive Mashco-Piro who is bathed in a black resin dye from head to toe. They stare at one another for a moment, then flee in opposite directions. Such an encounter is rare. For most who travel or live near the park's rivers and trails, the Mashco-Piro Indians exist, at most, only as muffled hoots or unseen eyes peering through the brush. Small abandoned campsites testify to the presence of perhaps a few hundred kinfolk. These two tribes, the Machiguenga and the Mashco-Piro, personify the two most common images outsiders have of Central and South America's rainforest residents.

Moser/Tayler/The Hutchison Library

A Tukano Indian dressed for a ceremonial dance in Colombia. Ritual ceremonies are essential to the maintenance of harmony within the Tukano community and between the community and the spirit world. Elaborate headdresses, such as the one pictured, are prized possessions of the Tukano and are generally kept in special woven boxes.

A Cultural Patchwork

In the minds of many outsiders the Mashco-Piro represent the romantic image: small clusters of closely linked kin, isolated from their neighbors and other strangers, and cloaked in mystery and the exotic. For others the image is that of the more visible Machiguenga, who are the sympathetic, esoteric subjects of Mario Vargas Llosa's novel, *The Storyteller*. They are regarded as objects for pity and charity as they become progressively marginalized victims, surviving within a protective and controlled park on the fringe of the dominant society.

Neither the romantic nor the tragic image does justice to the current range of tropical American rainforest Indian societies. Today no Indian group remains "uncontacted" by outsiders or unaffected by the national society. The Mashco-Piro are not hidden stone-age remnants; their grandparents—reflecting accurately on the rampages and atrocities of the rubber gatherers—cautioned them to be wary of outsiders. And their grandchildren paid attention. The Machiguenga enjoy the relative abundance of fish and game within the park, but they now rebel against rules laid down by administrators to "protect and preserve" Indian society. As true indigenes, the Machiguenga want to set their own limits and shape their own adaptation to a changing world.

Latin American rainforests and the adjacent upland savannah areas with forested riverbanks are home to about one million Indians, who are divided into approximately 300 ethnic or tribal groups. These Indians are very much part of their forest landscape. Several thousand settlements, ranging from small household clusters to villages of several hundred people, together with gardens of various sizes, dot the forests of the lowland Caribbean region of Central America, the northwestern coast of South America, and the vast Orinoco and Amazon Basins.

Michael Friedel

The forests' extraordinary biological diversity has housed similar cultural variety for thousands of years. Remnants of bones and stone tools suggest that hunters and gatherers moved through the understory and along the rivers at least 10,000 years ago. By 3000 BC and probably much earlier, Indians had domesticated a wide range of plants, produced agricultural surpluses, and settled into stable village life.

Today less than 1 percent of the forest groups of Latin America obtain most of their food by hunting and gathering. Despite popular images of forests laden with fruit and nuts, most foods are cultivated, and the rain-forest Indians depend on agriculture for subsistence and cash income. Along with occasional wage labor, the seasonal hunting and gathering of wild fruits and nuts is undertaken to supplement crop cultivation.

Gatherers and Nurturers

Historically and currently, these Indians have shaped the physical environment as well as adapted to it. With imagination, ingenuity, and intensive labor they have carefully managed and sustained their forest resources. They gather and nurture wild palms, fruits, or nuts. Smaller house gardens or isolated plantings produce flowers, teas, herbs, tobacco, fish poisons, curative vines or roots, and many other products amid the manioc, yams, and plantains that are the dominant staples. Indeed, rodents such as agoutis and pacas that ravage gardens, and birds that enjoy seasonal fruits live in such a symbiotic relationship with human settlements that their habits force a rethinking of the sharp distinctions often made between wild and domestic.

Indian fishermen on the Xingu River, Brazil. Obtaining fish with spears or by stunning them with plant poisons is common among many Amerindian peoples but no tribe relies entirely on fish for sustenance.

This map shows the approximate location of some of the peoples inhabiting the rainforests of Central and South America.

The rainforests of the Americas have also contributed significantly to the world's food supply. The Indians' ancestors domesticated a range of plants such as manioc and peanuts, and these are now major world food sources. Even today, indigenous use of wild species such as latex trees and curative plants has encouraged researchers—ranging from ethnobotanists to agronomists—to focus their attention on the Indians' vast botanical knowledge and techniques of plant manipulation.

Land is equally important to rainforest Indians. Now, as earlier, groups ranging from small, relatively isolated bands of Ecuadorian Huaorani or Bolivian Sirionó to large groups of Peruvian Shipibo and Ecuadorian Lowland Quichua shift their residence during the year. But none fits the ascribed stereotype of the perpetual nomad with little concern for property. Almost all Indians move in established patterns across familiar lands. Their concepts of land and resource rights may vary widely, but each group maintains clear rules which regulate their rights to land use, to garden plots and hunting trails, and to widely scattered fruit trees and medicinal plants. In short, natural resources are subject to indigenous conservation policies.

These rainforest societies illustrate social and cultural variety that defies any singular definition. Each rainforest group, in its own way, is becoming more visible and drawn into debates over land rights in the face of expansive national development. As recently as 20 years ago groups such as the 15,000 Yanomami of Brazil and Venezuela, the 1,000 Huaorani of Ecuador, or the several hundred Sirionó of Bolivia were still able to protect their territorial frontiers and maintain relative isolation through their ferocious reputations and occasional violence. The Tukanoans of Brazil and Colombia, the Quichua of Ecuador, and the Ashaninka of Peru—each with tens of thousands of kin—have maintained regular contact and trade with non-Indians for several centuries. Each of these groups now competes with colonists and industries for rights to land and resources. Some Indian groups have been quite successful. Leaders of the Kuna of Panama, the Shuar of Ecuador, and the Kayapó of Brazil shuttle in small private planes from their jungle homes to urban centers, where they negotiate rights to land, resources, and education with government officials and international organizations such as the World Bank. Most groups, however, are less fortunate.

In the course of protecting their land and conserving their resources, the tropical forest groups are also forging greater links with national societies. But just as each group has created unique world views, passed on or slightly transformed from generation to generation, so responses to the physical environment and reactions to external threats distinguish one indigenous group from another. The rainforest Indians' adaptation and response to their physical and social environment do, however, have some common historical and contemporary patterns.

Hunters and Fishers

Some things have changed little over thousands of years. For instance, scanning the rainforest canopy in search of spider monkeys or skimming the surface of a stream for fish stunned by poison are still among the Indians' most exciting and entertaining means of obtaining food. Hunting and fishing experiences also continue to inspire much conversation and story-telling. A hunter is the subject of scrutiny by the

Steve Cox/Camera Press London/Austral

A group of Kayapó in conference. A popular misconception of forest dwellers as mere foragers is especially inappropriate as applied to most of the peoples of Amazonia. The Kayapó, for example, are forest managers who use a wide range of sophisticated agricultural techniques, including crop rotation, seed selection, and soil fertilization. Their regular food supply is based on at least 250 species of fruit, nuts, tubers and leaves, and they select from some 650 medicinal plants.

entire community when he returns from a hunt, and his success or failure soon becomes public knowledge. Damming a river for a fish poisoning is a communal event that creates day-long excitement for women and children as well as men. Some Colombian Amazonian Indians even distinguish each other as either "fish people" or "animal people", in part linking the identification of each group's focus with its resource.

Yet animal biomass in the rainforests of Central and South America has always been small and widely dispersed. This was the case even before the recent deforestation and depletion of natural food resources. Wild game has always accounted for a relatively small part of the Indians' diet. Fishing, particularly in the nutrient-rich "white rivers," that drain alluvium from the Andean slopes, is a more reliable source of protein, but no society relies on fish for the bulk of its subsistence. Similarly, domesticated or wild fruits such as peach palm, papaya, pineapple, avocado, brazil nuts, and morete palm provide seasonal variety to the diet rather than being the principal food supply for an entire population.

Traditional Land Use

Because domestic plants provide the bulk of the Indians' nutrition, they are managed with greater regularity and predictability than any other food source. Rainforest groups generally maintain two planting areas. The first is the smaller "house garden", which abuts the dwelling, and is characterized by extensive variety rather than intensive production. Such gardens can be virtual cornucopias of decorative plants, minor food crops, teas, medicinal plants, condiments, and dyes.

Most of the major food crops are grown farther from the household, in larger gardens variously referred to as *milpas*, *chacras*, or *conucos*, and these plots range in size from 0.5 hectare to 1 hectare (1¼ to 2½ acres). Families usually maintain two or three plots in various forms of production and stages of regrowth. Foremost among the garden cultigens (plants known today only in a cultivated form) is manioc, a starchy tuber so efficient in terms of potential calories per acre that its only peers worldwide are plantains and rice.

These gardens also house a large number of other plants. What appears to the unfamiliar eye as a hodgepodge of cuttings, stalks, and vines frequently contains a carefully selected and well-ordered array of 50 or more types of food plant. Crops such as corn, beans, and peanuts are harvested first, whereas manioc, which spoils after a few days in the air, is depleted as needed. Once these primary crops have been removed, additional plantains are added to fill the space. When these are no longer productive, fruit trees continue to provide food and to mark the family's future-use rights to the plot. This cycle allows each garden to remain in some form of production for up to 30 years, while the clearing regenerates as forest. By contrast, current industrial use can so degrade soils that many rainforests are no longer usable after five to fifteen years.

For many years development specialists considered rainforest Indians' principal production system of swidden, or slash-and-burn, horticulture a primitive and cluttered example of agricultural backwardness that should be replaced with intensive monoculture farms. However, detailed study during the past 20 years has revealed much complexity and planning in swidden practices that reflect sophisticated plant knowledge and creative management. In fact polycultural swidden horticulture is not only extremely productive, it is perhaps, given their soil quality, the rainforests' most ecologically well-adapted and renewable method of land use.

It is small wonder that skilled resource managers should have developed so sustainable an agricultural system. However, in the recent surge of concern with tropical deforestation, many environmentalists seem to have overlooked these findings. Slash-and-burn horticulture is often named as a major cause of tropical deforestation, and Indians, by extension, are cast as agents of destruction. Yet, swidden has been practiced for

Below. *Small boys everywhere have pets, whether furred or feathered: an Uru Eu Wau Wau boy takes his pet macaw for a walk in the forest, Rondonia, Brazil.*

Center. *Supervised by his pet monkey, a young Xingu boy digs for sweet potato. This and plantain constitute the staple diet of many Amerindian tribes in Amazonia, supplemented by nuts, fruits, and berries from several hundred rainforest plants.*

Loren McIntyre

Claire Leimbach

centuries, which is evidence that it is a sustainable system. Confusion arises when the indigenous horticultural approach is lumped in with large-scale land-use modifications—agribusiness and cattle raising—which employ some of the same practices.

Though Indians may slash and burn the forest, they do so in a manner that creates small gaps laid out as isolated patches. Once such a sector has been opened by felling the trees and clearing the brush, the litter is left to dry before it is burned or, in those areas that have no real dry season, to decompose as mulch. Rapid-growing crops such as corn are often planted first, before the garden is cleared for intensive production of manioc or plantains and a wide range of other tubers and seed crops. After about 30 years secondary forest dominates, thorny brush disappears, and while the agricultural cycle is begun anew at another site the old one is reabsorbed into the forest. Even small gaps opened in the forest canopy by naturally falling trees affect the diversity of regrowth, and swidden plots simply mimic these natural forest dynamics, although on a slightly larger scale. Regrowth may alter the earlier structure somewhat, but the surrounding woodlands facilitate natural regeneration with little alteration to the initial biological diversity.

By contrast, cattle ranches and agribusiness plantations are established by slashing and burning areas so great in size that there can be no mimicry of the cycles of either swidden or natural reforestation. Large clear-felled areas do not lead to biologically diverse regrowth, and as a result their modifications to the landscape are significant and generally permanent. Such intrusions are undoubtedly the greatest threat to the forests' social and biological future.

A Kuna woman, machete in hand, returns homeward with a basket of produce from her forest garden in Panama. The Kuna depend largely on slash-and-burn agriculture, supplemented with fish and small game, much as they have since the time of first European contact in the sixteenth century.

Danny Lehman

Victor Englebert/Time-Life Books

Unfortunately, confusion of indigenous land practices with truly destructive systems has sometimes led to misunderstandings between rainforest Indians and environmentalists who should be allies. Meanwhile, extraction and intensive production systems by non-Indian people present perhaps the greatest dilemma and greatest challenge for both the Indians and others concerned with the future of rainforests and rainforest societies.

First Contacts

Before 1492 there were perhaps 7 to 10 million Indians living in the rainforests of Central and South America. The communities in which they lived were radically different from those of today's rainforest Indians, in terms of political organization and wealth as well as numbers. In fact, early Central and South American rainforest residents relied on access to rivers and riverine resources—as avenues for movement, as zones of alluvium for intensive agriculture, and as sources of fish and protein—rather than on the forest. Archeological evidence suggests that riverine resources were critical to the growth and expansion of these societies, and there is a direct relationship between the richness of material life in excavated settlements and their proximity to water and to fish protein. Sites along rivers were large and produced elegant, carefully manufactured material goods, and these were in sharp contrast to the poverty of remains from excavations of settlements away from the river. There is also evidence of an ebb and flow of groups, identified by their specific design motifs on ceramics, between highly desirable riverine sites and inland settlements. This suggests a continuous struggle—groups gaining and then losing territory and access to critical resources.

Within 100 years of first contact with Europeans, the Indian populations were reduced by up to 90 percent and the survivors dispersed. Ironically, it was those large, well-organized groups in river settlements that suffered the most. They were popular targets for slavery and were also the most vulnerable to deadly epidemics of introduced diseases including smallpox, measles, whooping cough, chicken pox, bubonic plague, typhus, amebic dysentery, and influenza. Most of these were endemic in Europe at the time, but they devastated unexposed populations in the Americas and the epidemic pattern has persisted to the present.

With health concerns still weighing heavily on them, Indians of the American rainforests react to disease with good sense and historical hindsight. The safest refuge is relative isolation, and even the most densely populated communities usually maintain a settlement pattern that allows them to disperse when epidemics threaten. Villagers describe numerous and recent incidents in which an individual shows signs of an illness such as whooping cough, and by the following morning the settlement is

Victor Englebert/Time-Life Books

abandoned, its entire population having fled to secondary houses scattered deep in the interior. These sites often offer better hunting and fishing, but they are also essential refuge areas.

Populations that were dispersed and away from the rivers survived European invasion and colonization at a greater rate than those who relied on the river, and their lifestyles and settlement patterns were subsequently adopted by many other Indians. So, despite catastrophic social and demographic destruction and upheaval, rainforest Indians have not only survived, but today they constitute one of the fastest growing segments of Central and South America's population.

However, these groups have no further refuge from disease or from non-Indian encroachment. The interfluvial forests are the new frontiers for both colonists and exploitative industries. Together they cleave off the edges of forests and slice roads through the woodlands' interior. Chainsaws and bulldozers then begin clearing areas that were once inaccessible. This advancing frontier currently threatens the shape and even the existence of Latin America's heterogeneous social and physical mosaic. But unlike the forests, the human residents can challenge the forces of change.

Ethnic Federations

In the face of invasive cultures and industries, American rainforest Indians often adopted lifestyles in which they either mimicked or subordinated themselves to those who usurped their lands and resources. But since the second great wave of depopulation (when rubber gatherers spread throughout the rainforests) in the late nineteenth century, and now faced with losing their last lands, the indigenous response has changed.

Indians have decided to protect themselves and their resources. At a local, regional, and international level their leaders are working to regain stewardship of the forests in ways that parallel their earlier adaptation—that is, not as passive objects of environmental or social determinism, but as both resource managers and architects of their social status. Indian social mobilization is not limited to the Central and South American rainforest, but it is here that unique organizations—ethnic federations—have attained their most highly organized and culturally distinct forms. These federations are now one of the defining features of contemporary forest landscape and must be heeded by all those who hope to either develop the environment, or conserve it.

A group of Yanomami pause to rest during a food-gathering expedition. The Yanomami are unusual among Amerindian societies in their dependence on plantain (a close cousin of the banana) as a staple crop. It accounts for about 70 percent of their diet, which is usually supplemented by a wide variety of wild plants gathered during frequent excursions into the forest.

Opposite. *Confronting an uncertain future, a young Yanomami stands, machete at the ready, as though in symbolic guard duty over his homeland. The Yanomami inhabit the northern borders of Brazil, a "wild west" region of violence and lawlessness. In pursuance of a major national security project (the Calha Norte Plan), the Brazilian military has moved into the area in force, and at the same time the region has been overrun with small-scale gold miners. Each side fights the other, and both fight the Yanomami.*

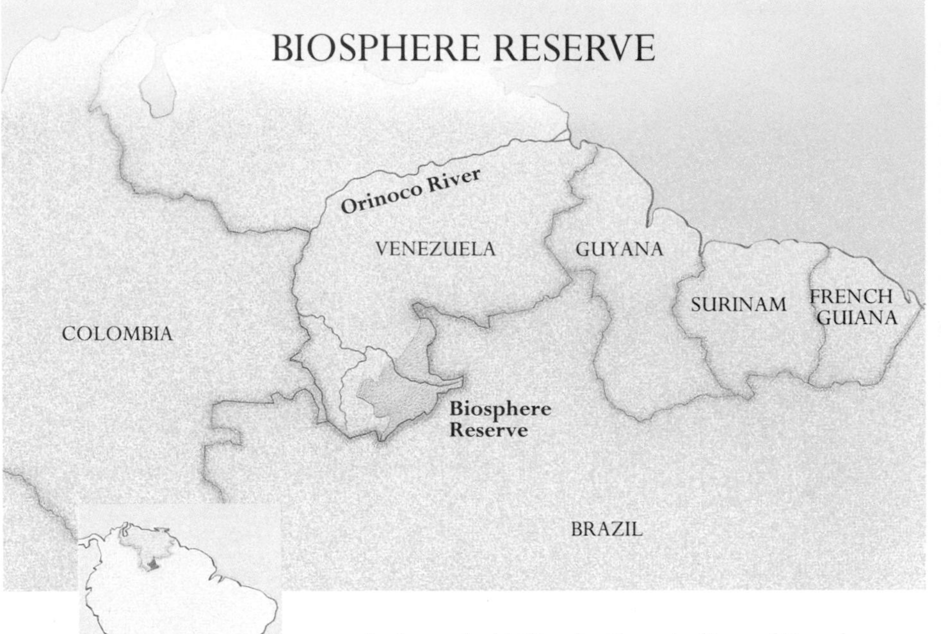

During 1991, the headwaters area of the Orinoco River in southern Venezuela was set aside as the Orinoco-Casiquiare Biosphere Reserve. Sprawling over 83,000 square kilometers (32,000 square miles) of rainforest, it is the largest reserve of its kind in the world. It is also the home of the Yanomami and Yekuana Amerindian tribes, and although this vast area was set aside largely as a concession to the land rights of these indigenous peoples, its establishment was by presidential decree, and has not yet been given the force of law.

Opposite. *Elaborate body paint and partly shaven heads characterize a group of Kayapó children. The paint consists of mixtures of charcoal and the juice of various fruits. Bold stripes are a common design feature, suggesting inspiration from the markings of bees and wasps. Perhaps it is significant that the Kayapó believe their ancestors learned their communal lifestyle from such social insects.*

In the early 1960s, the Shuar Indians of Ecuador formed the Federation of Centros Shuar to oppose colonists and development programs that were spilling over the Andes and into the upper Amazon. Shortly thereafter, the Amuesha Indians of Peru's central jungle established the Amuesha Congress. Over the past 20 years similar organizations and programs to resolve common problems have proliferated. The movement, though political in form, is distinctly ethnic in terms of aspirations. Unlike groups acting strictly from a standpoint of class or economic status, rainforest Indians are reflecting on and drawing from their unique cultural and environmental history and experience. Their organizations' main concerns lie with their empowerment (in the face of encroaching non-Indians with their local and national programs), maintenance of ethnic identity, and control of both rights to and use of their land and resources.

For most rainforest Indians of the Americas, talk of "fragile ecosystems" connotes a political as well as physical environment. Although they are deeply involved in protecting and managing their resources, most of them are not conservationists from a standpoint of global philosophy. They are pragmatists who recognize that their land and its resources are their present and future capital. In the past, Spanish *conquistadores* and *encomenderos* robbed them of their land and labor in the name of the crown and in exchange for the "word of God". More recently, development and conservation planners have designed land-use programs on Indian lands in the name of world environmental strategies and in exchange for management plans, without actually consulting the Indian "beneficiaries". As a consequence Indians are now wary of any proposed form of development or conservation. "Development or conservation for whom?" they ask, and with good historical reasons for doing so.

Steve Cox/ Camera Press London/Austral

Jono Ripper/D. Donne Bryant/Stock Photos

Portrait of a young Kayapó girl. This type of facial decoration is common among many of the indigenous Amerindian tribes of Amazonia.

Similar questions regarding any proposed activities can be expected from Indians and other forest residents, particularly when the project affects their land and resources. Most development agencies as well as environmental and human rights organizations, be they national or international, rarely take the time to understand the broad political and historical concerns of rainforest Indians, who are often regarded as passive victims—another "species" to protect from extinction or to aid through development. Although essentially sympathetic, such attitudes exclude the residents from participating in most decisions and neglect the priorities of the many Indian organizations that now link the patchwork of communities in the Central and South American rainforests.

The federations' concern is not just over the management of their land and its resources, but for permanent and secure access to them. They are fully aware of the implications of statements such as the Declaration on Environmental Policy and Procedures, and Procedures Affecting Economic Development, which was signed by 10 bilateral and multilateral development banks in 1979. While the document makes strong recommendations in support of environmental issues, it also recognizes the right of national governments to determine priorities and forms of national development. Consequently, Indians who have seldom participated in the government planning that most affects them fear that their capital base stands greater risk of loss than does their understanding of resource management. So land tenure is awarded considerably more public attention than its management.

Faced with broad regional problems, federations in the Amazon basin have also begun to organize internationally. From 1984 onwards, ethnic federations from the rainforest regions of Brazil, Bolivia, Peru, Ecuador, and Colombia joined to form the Coordinating Body for Indian People's Organizations of the Amazon Basin (COICA). Originally established to give voice to groups rarely heard in international forums like the United Nations, COICA now has leaders who speak directly to national governments and to officials of multilateral lending agencies such as the World Bank and the Inter-American Development Bank. In 1988, COICA representatives traveled to Washington DC to inform the international environmental community of Indian needs and concerns. Subsequently, COICA convened an international summit in Iquitos, Peru, in an endeavor to seek ways of protecting both the rainforests and their Indian residents through alliances with environmental organizations. This collaborative process still continues.

Ethnic federations are also unique social mechanisms for present and future forest management. Sparked by efforts such as the Kuna Indians' conservation and land-rights program—Project Pemasky—numerous ethnic federations have now established broad resource-management programs. These focus on management, but they are also linked to land and resource rights and seek to maintain the unique ethnic identities of indigenous groups caught up in social change.

Danny Lehman

Accordingly, Indian organizations and the programs they have created do not always evaluate the benefits of a particular activity by the technical or economic criteria used by development or environmental planners. Rather, they strongly weigh its relation to their own larger goals—empowerment of their organizations, land rights, and maintenance of a strong sense of self. Such movements by the Indians of Central and South America, and respect

for their needs by the environmental community, are beginning to demonstrate some promising signs amid apocalyptic cries of ecological destruction.

To reverse the current trends toward the elimination of the Amazonian and other tropical rainforests of Central and South America, Indian collaboration is essential. If Indian organizations are excluded, the story of Sinaá, the culture-hero of the Juruna of Brazil, seems particularly appropriate: "Sinaá first made the earth a livable and safe place. And then he went to live downriver. Sinaá goes on living there downstream where he had gone. A long time ago a Juruna visited him... after Sinaá asked how his people were, he took his guest up to the top of a large rock, from which the Juruna could be seen down below, fishing in their canoes. Finally Sinaá showed the Juruna visitor an enormous forked stick that supported the sky and said, 'The day our people die out entirely, I will pull this down, and the sky will collapse, and all people will disappear. That will be the end of everything.'"

Rainforest Indians are fully aware of the use and power of these mythical metaphors. If the current social mobilization of Indian people and the future alliances this movement can engender among environmentalists move forward, the prophecy of Sinaá may not become the reality many have predicted.■

Kuna women commuting between their coral island homes off the east coast of Panama and their farms on the mainland. Unlike other indigenous populations, the Kuna own the land they live on, and in the 1970s they negotiated a range of land-rights, which resulted in the Kuna Wildlands Project, a scheme designed to protect the Kuna and their lands from outside interference. The success of the project has made it a model for other indigenous peoples in Central and South America.

8 The Impact of Europeans

MICHAEL WILLIAMS

Before AD 1500 civilization was essentially land-centered. Contacts by sea were relatively unimportant, and, except for a few overland trading routes linking Europe with India and Africa, the continents remained isolated from one another. But the discoveries of Christopher Columbus and others after 1492 established sea contacts between continents and started the process that led to Europe's cultural and economic domination over much of the globe.

E.T. Archive

The emergence of tea and coffee as popular beverages in Europe during the seventeenth century brought with it an enormously increased demand for sugar. This in turn led to the earliest phase of the global destruction of rainforests, as large areas were cleared for the establishment of sugar-cane plantations as a profitable cash crop, a development symbolized by this contemporary engraving of a Negro slave cutting cane in the Caribbean in 1799.

Alien Forests

Before the voyages of discovery, Europe was a peripheral appendage of the civilized world that consisted of the land-based empires of Ming China, Ottoman Middle East and North Africa, Safavid Persia, and Mughal northern India. After about 1500 Europe gradually moved to being at the center of world innovation, trade, and change, and became the most powerful region in the world. With this shift in focus and expansion of European influence, the harmonious relationship between the inhabitants of the world's tropical forests and their environment was shattered.

Of course, all the forests of the globe were to be affected by the expansion of European influence. In the temperate woodlands of eastern North America and eastern Russia, for example, the forests were familiar and therefore were a tempting stage for the re-enactment of the great European colonizations of the early Middle Ages, which had seen vast areas of mixed hardwood forest cleared for agricultural settlement.

But the forests of the tropical world were inhospitable and strange, dominated by alien cultures, and there was little permanent European settlement or clearing for agriculture before 1800. Except in the Caribbean islands and Brazil, settlement was mainly coastal and strategic, focusing on guarding sea routes and anchorages and on controlling interior trade in the hinterlands. In time, however, even this marginal settlement affected traditional cultures and societies. They were drawn into the European network of world trade and in the process they were changed.

The Real Impact of Discovery

After Europe became conscious of the wider world its trade moved from its medieval preoccupation with small quantities of high-value luxury goods such as spices, perfumes, porcelain, dyestuffs, silk clothes, and rugs from Asia, and slaves, ivory, and gold from Africa, all of which were capable of withstanding the cost and rigors of long overland transport. In their place arose a trade in goods that in time developed into a mass trade for an increasingly affluent population.

Rijksmuseum Amsterdam

Many of these commodities, such as tea, coffee, sugar, potatoes, and cotton, were tropical products that related to basic needs of food and clothing. Others, such as tobacco, were more optional. Initially all these goods were regarded as exotic products that only the wealthy could afford but gradually, with mass production methods of manufacture and distribution, they became cheaper and within the reach of more and more people. With this transition, the impact on the tropical forests really began, particularly as Europeans had the ships and seafaring knowledge to move plants and animals around the world, either to where they grew best or to where production could best be controlled for imperial needs. The breaking down of the biological isolation of the continents that had existed for millennia and the replacement of one vegetation cover by another were supreme examples of what the voyages of discovery really meant.

Individual trees, or at best scattered plots, that existed previously were concentrated in new units of production: plantations. These concentrated on one crop and maximized output. If labor was not available in

Although the Portuguese were the predominant colonists of Brazil, the Dutch held a small area for several decades in the mid-seventeenth century. During this time, painters such as Albert Eckhout, who painted this work entitled A Market Stall in the Indies, *were commissioned to visually document the colony.*

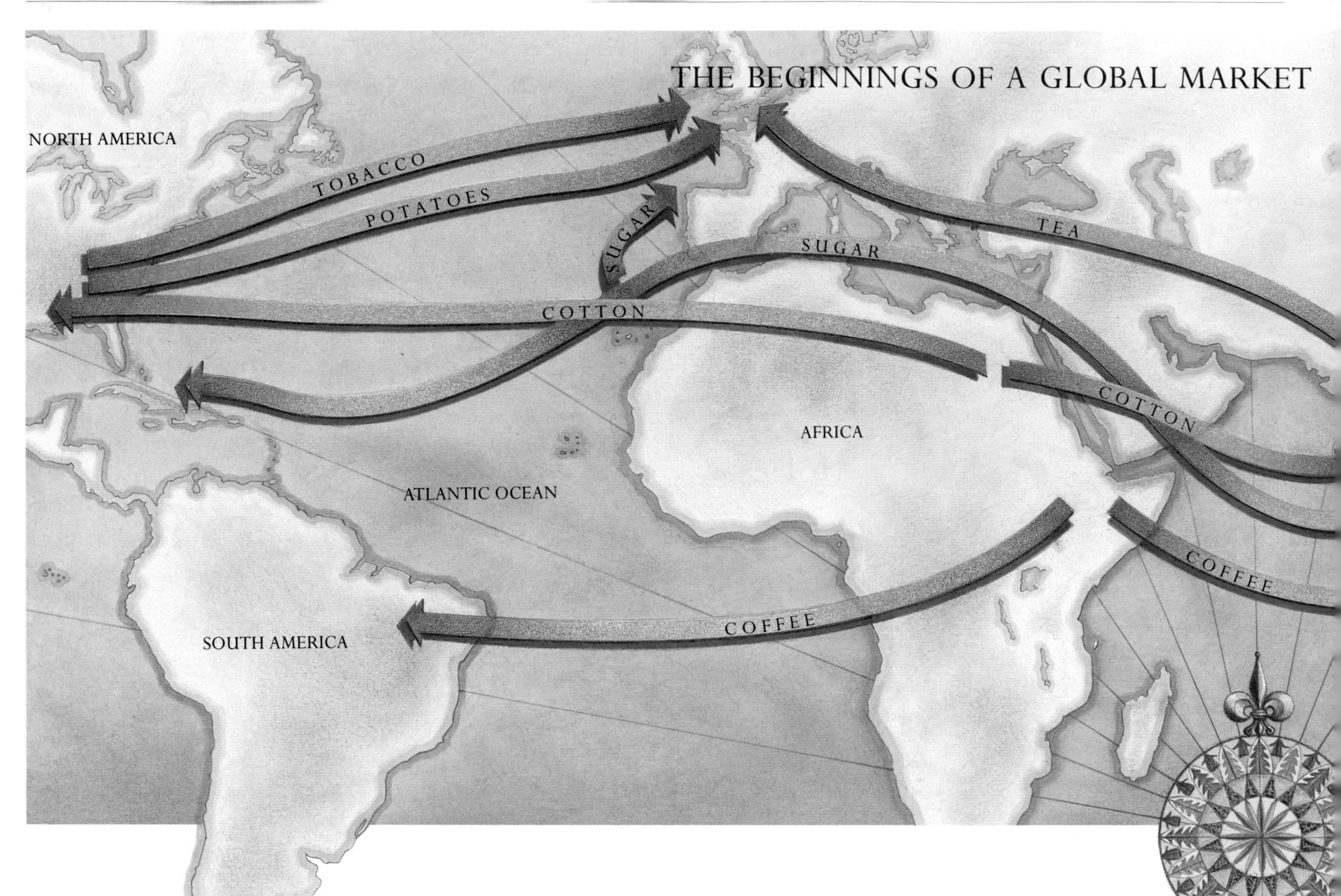

A fiber as old as Egyptian civilization, cotton was indigenous to most of the tropical and subtropical world. With the invention of mass production methods in England during the late eighteenth and early nineteenth centuries, cheap, easily washable cotton underwear and clothing, which were seen to be hygienic and fashionable, became available to many people.

The potato was another crop from the New World. It took a long time to be accepted as a useful food product, but by the end of the eighteenth century whole sections of the population in Europe, usually the poorer, and whole societies such as in Ireland, were subsisting on them.

Unlike many other crops, tobacco was a New World product that was moved to the Old World. Originally considered a medicinal cure for just about every ill, tobacco was not commonly used as a stimulant until the early eighteenth century, and it only became a mass-consumption commodity a century later with the invention and factory production of cigarettes.

Tea was introduced to Europeans early in the seventeenth century, about the same time as coffee, by merchants who imported it from the Orient. Initially, it was regarded as a luxury and heavily taxed. However, by the mid-eighteenth century it had become a fairly common drink.

Coffee probably originated in Ethiopia and was moved to Indonesia by the Dutch about 1650 and then to Brazil by the British around 1700. It was introduced to Europe around 1650, the new institution of the coffee house both reflecting and promoting its acceptance.

Sugar cane probably originated in the wetter regions of Asia or the south Pacific, and although its product (sucrose) had been introduced into Europe as early as the first century AD by Arab traders, it did not become an indispensable element of Western diet until it could be refined and became associated with the growing cults of tea and coffee drinking.

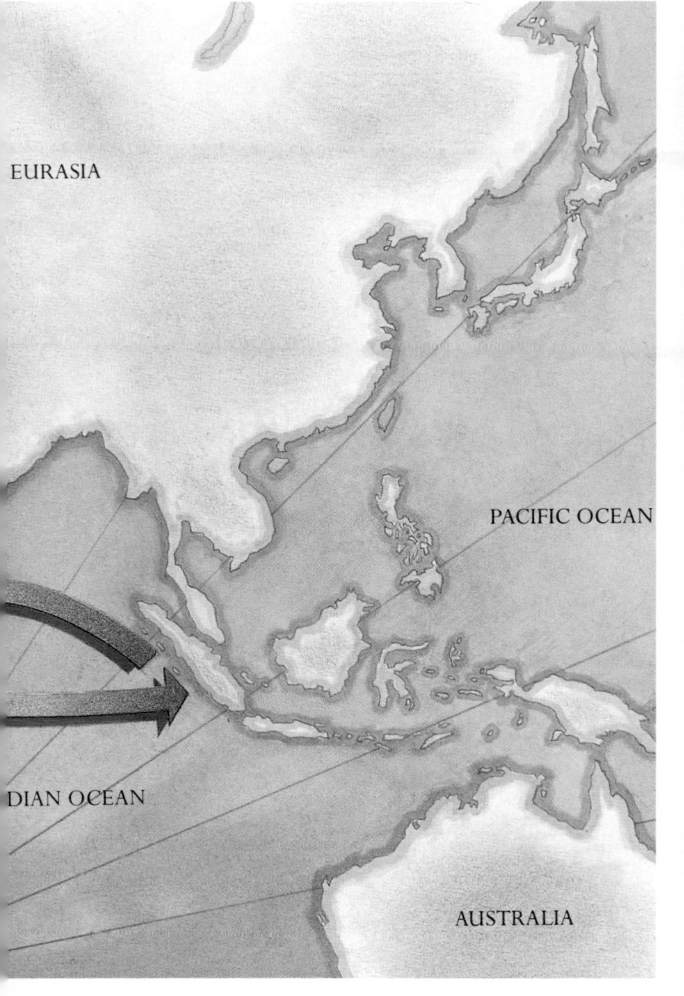

Coo-ee Historical Picture Library

A woman plucks the new shoots from a tea bush on a Sri Lankan plantation, c. 1905. Tea is traditionally plucked by hand and an experienced plucker can harvest about 18 kilograms (40 pounds) of tea per day.

the quantities needed for this new scale of production, it too was moved around the world. Slaves were captured in Africa and transported to the Caribbean, Latin America, and the southern United States. Later, indentured laborers from India were hired and taken to sugar plantations in Fiji and Guiana, rubber plantations in Malaysia, and tea plantations in Sri Lanka and Natal. It was not only the plants and people that were moved. Goats, sheep, horses, pigs, mules, and cattle were introduced into the New World and later to Australia and New Zealand—always with devastating effect. The stocking of these grazing animals encouraged clearing and pasture development, and it inhibited the regeneration of young trees. With their hard hooves the stock trampled native grasses, increasing soil erosion and aiding the spread of Old World crops and weeds.

Even if the tropical forests were not directly and actively settled and felled by Europeans or European overseer entrepreneurs, the European impact still impinged on peasant agriculturalists. By the end of the eighteenth century European traders such as the East India Company became very active European colonizers, annexing large chunks of territory so they could gain complete economic and political control of resources. In that process they usually instituted a system of land apportionment, survey, and taxation. Since taxes usually had to be paid in cash, peasants began to grow cash crops, thereby entering the global economy. In so doing they felled yet more forest in order to produce above subsistence level.

In brief, the European voyages of discovery began a biological and commercial invasion of the wider world

Tropical rainforests captured the imagination of eighteenth and nineteenth century Europeans as something strange, ancient, and utterly indestructible, a sort of living cathedral—a concept captured in this painting by Charles, Comte de Clarac (1877-1927). Only in the twentieth century have we had to come to terms with the innate fragility of this ecosystem.

that was eventually to have as much of an impact on the tropical forests as did the better-known European immigrant invasion of the forests of the temperate world. Whether these new foods and clothing and their greater availability in Europe contributed to better nutrition, hygiene, and health—and hence to population growth—is an open question. Nevertheless, there was a steady increase of population during these centuries. By the early nineteenth century the increase had become so great that some 50 million Europeans were able to migrate overseas during the next 100 years, most, to be sure, to the new temperate territories, but many to the tropical world.

A Very Different World

The initial contact of Europeans with the rainforest brought them face to face with societies and methods of land use that varied enormously. In Latin America, Central America, the Caribbean, and the southern portions of North America they found vast areas of forest inhabited by few humans, most of whom were hunters and gatherers. Elsewhere, however, there were patches of intensely cultivated land that resulted from forms of swidden, or "slash-and-burn", agriculture.

This form of managing the land meant that the forest, the secondary bush, and woodland vegetation were cleared with simple hand tools. Trees and shrubs that either bore favorite fruits or were useful for medicinal purposes were often left standing, as were larger trees that were not worth the effort of felling. Others were pruned down to stumps. The cultivation phase alternated with much longer periods of fallow that lasted up to 20 years, during which time the vegetation regenerated naturally, invaded the plot, and obliterated it. Because the Amerindians had no livestock they had no interest in establishing pastures. Consequently, when agricultural clearings were abandoned they were not burned over to encourage grass growth. Instead, they were allowed to revert directly to a secondary forest that was often dense, though never quite the same as the original forest in its species composition. Nevertheless, a modicum of nutrients were restored to the soil. It was much the same in Africa.

In the tropical forests of southern Asia, Europeans found well-established economies and centuries-old societies in which cultivation had reached a high art. The tropical forests, especially the dry tropical forests, had long since been felled and replaced by intensive cultivation, of paddy rice in particular, often on terraces with irrigation and drainage systems that were complex and labor-intensive in the extreme. Other crops which were later to be so attractive to Europeans included coconuts and tea.

The contact with explorers, missionaries, and colonists had a number of effects. First, in the simpler, more consumption-orientated economies the primitive stone axes were replaced by more efficient steel axes and machetes. These did not significantly alter the use of the forest. The lives of the people and their daily habits were still intertwined with the annual ritual of forest clearing, planting, fallowing, and moving on, although steel tools did speed up the process considerably. But what was devastating was the unintentional transference of relatively simple European diseases such as measles and influenza—and smallpox after 1518—that devastated the Amerindian populations.

After the initial probes of explorers, groups of Europeans intent on serious settlement landed and found large areas of cultivated land, particularly maize fields, bereft of people. The perception was that these were god-given prizes for the new settlers, when in fact they had only recently been depopulated. Later despoliation and conquest extinguished the last remains of these once-glorious civilizations and pushed the remainder into the backwoods.

These generalizations about European impacts are best understood by looking at specific examples of what happened in the tropical world before 1800.

THE CARIBBEAN

Some of the earliest European contacts with the tropical forests were in the islands of the Caribbean and on the nearby Mexican mainland. The native Caribs and Arawaks impressed the Spanish with the sophistication of their agriculture, pottery-making, use of fibers, and canoe construction, and a social structure in which they seemed to live in quiet and peaceful harmony with themselves and their environment. Nevertheless, the Amerindians quickly succumbed to the Europeans' superior force of arms, the cultural invasion of their homelands, and the associated introduction of Old World microorganisms to which they had little resistance. Forced labor and slavery quickly finished them off so that few were left by the middle of the sixteenth century. It was to be a recurrent story.

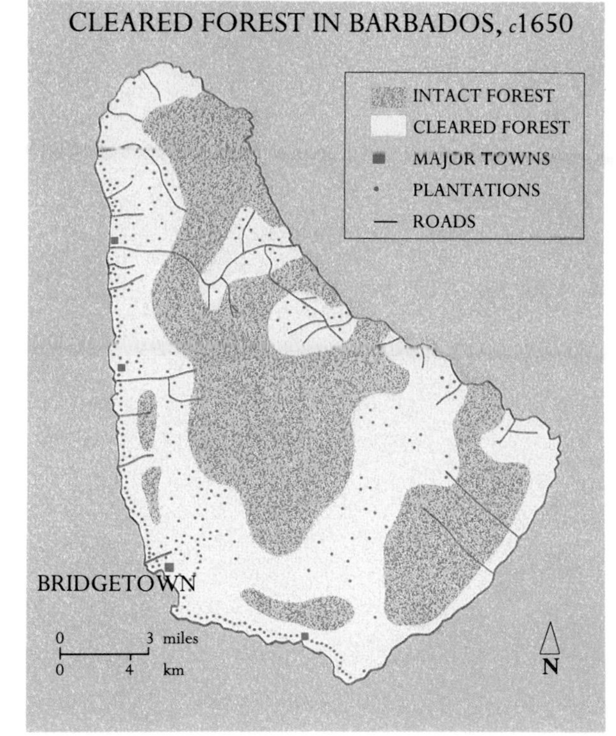

Islands in the Caribbean were the first of the rainforest areas of the world to fall to the European axe. The ensuing destruction has been especially well documented in the case of Barbados, where forest was systematically cleared for commercial sugar-cane plantations within a few decades of its discovery. The map shows the situation around 1650.

With the elimination of the native population and the abandonment of cultivation, it is possible that at first the forests actually expanded in size and density. But the colonizers' desire to establish an economy with an export staple led to the introduction of sugar cane from the Canary Islands to Esponala (Haiti/Dominica) by as early as 1498. Bananas followed in 1512, together with other Old World crops and animals, particularly pigs, which bred rapidly and ran wild in the forests.

Generally, the Spanish were transitory plunderers of wealth. They brutally exploited the populations in their search for gold but paid relatively little attention to settlement. The Dutch and English who followed in the sixteenth century were often intent on more permanent settlement. By 1642 Barbados and St Kitts had a settler population of around 60,000, which represented about 1 percent of the British population. Many were sponsored and encouraged to go there by the new companies that were attempting to settle the land and stimulate trade, and that would soon be introducing black African slaves to their plantations.

The record of the impact on the forests is particularly clear in Barbados, where after 1625 the overriding aim was to plant sugar cane as a commercial crop. Initially small patches of forest were selected and removed by axing. Stumps were left to rot in the ground while the harder-to-cut large trees were left untouched until settlers learnt the native technique of

Before the period of European expansion, crops were grown mainly for food, and any foreign settlement was associated mainly with garrison support. But merchants came to see that large profits could be reaped from the importation of exotic goods into the European market, and thus the "cash-crop" was born. Plantations were quickly established purely for profit, involving crops as diverse as bananas in Panama and tea in Sri Lanka.

Opposite. *A contemporary view of Negroes at play in the early eighteenth century in the New World. The rapid and widespread establishment of plantations in tropical America in the sixteenth and seventeenth centuries resulted not only in local destruction of rainforest, but also a vigorous expansion of the infamous slave trade—a forced migration of huge numbers of people to serve as a labor force, mainly involving the indigenous peoples of Africa.*

Coo-ee Historical Picture Library

clearing by first ringbarking and then burning the dried-out tree towards the end of the dry season.

Soon, systematic clearing replaced selective clearing. This led inexorably toward the total destruction of the rainforest and the secondary scrub along the coasts. By 1647 about 20 percent of the timber on Barbados had been cleared and extensive sugar plantations established.

But the worst was still to come. Large inputs of capital and the importation of black slaves from Africa ensured the profitability of the plantations, and they expanded rapidly, consuming the bulk of the remainder of the forests. By 1665 the once forest-covered landscape of Barbados was almost totally open and dominated by large sugar estates.

By 1672 the same devastation and transformation had overtaken the neighboring islands of St Kitts, Nevis, and Montserrat. By 1690 the same was true of Martinique and Guadeloupe, and Jamaica was not far behind. The severity of clearing throughout the Caribbean was highlighted by the growing scarcity and high price of timber for construction and fuel wood, particularly for refining the sugar, so that by 1671 it was reported that: "at the Barbadoes all the trees are destroyed, so that wanting wood to boyle their sugar, they are forced to send for coales from England."

The alien, commercial objectives of Europeans serving a worldwide market resulted not only in cleared forests but also a whole host of further ecological and biogeographical changes. Native fauna and flora disappeared and were replaced by opportunistic and aggressive Old World varieties. Ferns, thistles, nettles, artichokes, and plantain, as well as the more common grasses and clovers, found eco-niches that had never existed before on the bared and eroded ground, and they took over. In other places the humus layer was leached or destroyed with detrimental long-term effects. Wild pigs and cattle replaced monkeys.

Although not very extensive, the islands of the Caribbean experienced intense change in their confined areas in a remarkably short time. They were a microcosm of the European impact on the tropical forests of the rest of the world, and for the rest of time.

BRAZIL

In contrast to the relatively small area of forests in the Caribbean islands, those of Brazil were among the most extensive in the world. In addition to the well-known Amazonian rainforest there was once a vast subtropical rainforest of approximately 780,000 square kilometers (300,000 square miles) that stretched along the Atlantic coast from Recife (Pernambuco) in the north nearly to Pôrto Alegre in the south, through the states of Pernambuco, Algoas, Sergipe, Espirito Santo, Rio de Janeiro, and most of the eastern portions of Bahia, Minas Gerais, and São Paulo. It was closed forest with a rich variety of species of tropical hardwood trees, such as ironwood and brazil wood, which were entwined with lianas and epiphytes. On the humid coastal-facing escarpments it closely resembled the great Amazonian rainforest to the northwest.

In pre-Columbian and early European times Amerindians practicing swidden agriculture grew mainly manioc, but also some corn, squash, beans, peppers, and peanuts. When European contact intensified after 1600 the native population that had survived disease was killed, driven out, or enslaved in much the same way as in the Caribbean. In their place a mixed-race, or *mestizo*, population emerged that adopted Amerindian agricultural methods. However, the swidden was intensified by the introduction of iron axes, machetes, and pigs.

Once fertility decreased, the farmers shifted on into the abundant forest and repeated the process. The

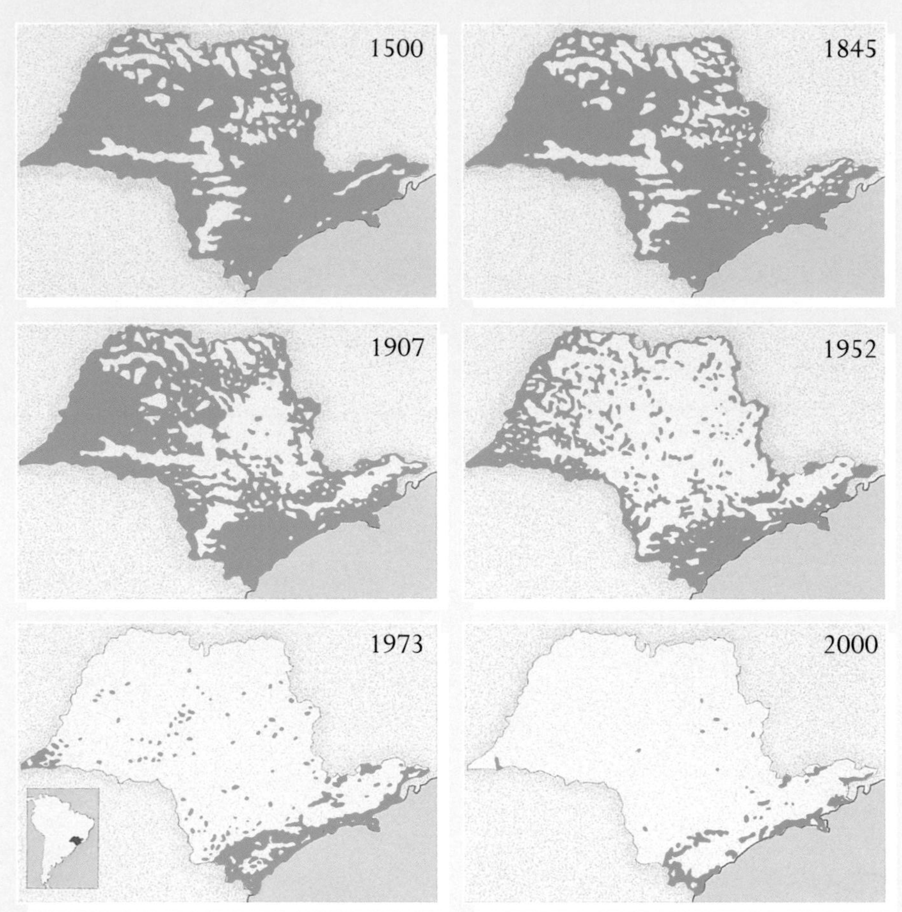

At the beginning of the sixteenth century, an estimated 81.8 percent of the total land area of the Brazilian state of São Paulo was forested. Land was cleared slowly at first, but the pace accelerated rapidly after 1860, with the proliferation of coffee plantations and the advent of railroads. Today, the original forest has been almost totally destroyed, and São Paulo is among the most heavily populated and industrialized regions of Brazil.

forest did not regenerate under such widespread cutting and burning, and the open ground was colonized by exotic grasses, ferns, and weeds, the process of change intensified by cattle and horse grazing and foraging by pigs. As in most pioneer societies, cleared land came to be valued more highly than forested land because it was "improved". And because the system was condoned it spread rapidly. Consequently, white and *mestizo* agriculture was the first step in the establishment of export-oriented plantation crops.

Although intermittent logging of the valuable hardwoods caused some local destruction, it was minor compared with the impact of gold and diamond mining. In 1690, the discovery of gold in Minas Gerais, and scattered pockets of gold and diamonds elsewhere, led to a further inflow of migrants and more forest destruction, so that possibly up to 20,000 square kilometers (about 7,500 square miles) were felled. The food and the fuel needed to support the new economy led to further great swathes being cut through the forest as farms and ranches were formed. It is estimated that another 25,000 square kilometers (almost 10,000 square miles) were cleared for planting manioc, maize, and rice, with larger areas of perhaps twice that size cleared for ranches, making a total of 95,000 square kilometers (about 37,000 square miles) cleared in the 100-year life of the mineral workings.

By the mid-eighteenth century the long-established cultivation of the main export crop, sugar, was putting pressure on the forest. Sugar had been introduced from Madeira in about 1560 and cultivation had been concentrated in the narrow and intermittent coastal strip from a little north of Pernambuco to just south of Bahia, with a minor concentration around Rio de Janeiro and the Campos region. Primary forest was always favored for sugar growing, and the clearing and harvesting were done by slaves. As extreme heat was required to crystallize the juice, vast additional areas were cut for wood fuel.

ASIA

Elsewhere at this time, the purposeful alteration of the tropical forests by Europeans was marginal in comparison with their impact in the Caribbean, Central America, and Brazil. The well-established and densely populated societies of southern Asia kept Europeans at arm's length, and China and Japan resisted incursion until the mid-nineteenth century.

Even though the direct European impact in these regions was slight, the traditional societies were not necessarily any more homogeneous, egalitarian, or caring of their forest resources. Studies of southwest India and Hunan province in China from the sixteenth century onwards have, for example, shown that the commercialization of the forest was not a European invention. In southwest India permanent agricultural settlement and shifting cultivation existed side by side, and village councils regulated how much forest exploitation could be undertaken by agriculturalists. But the forest was not regarded as a community resource and larger landowners dominated its use in their localities. Scarce commodities such as sandalwood, ivory, cinnamon, and pepper were under state or royal control. In Hunan, a highly centralized administration encouraged land clearance in an effort to increase both the tax base and local state revenues in order to support a bigger bureaucracy and militia. In simple terms, tropical forests were being exploited and diminishing in size as a response to increasing population and complexity of society.

After the Industrial Revolution

The European enmeshment of the economies of the tropical world intensified throughout the seventeenth and eighteenth centuries, and received a great boost from about 1750 onwards because of the changes associated with the Industrial Revolution. With this growing economic power came political power as the last unclaimed pieces of land were absorbed into the

REFLEJO/Susan Griggs Agency

An irrigated banana plantation in Ecuador. One consequence of European exploration and expansion was the wholesale translocation of a range of food plants around the world. Bananas, for example, originated in Asia, but are now one of the mainstays of the Ecuadorian economy.

new world system of colonial territories focused on Europe. The extension of shipping (especially steam) and railways brought standards of regularity, speed, and cheapness to passenger and goods traffic, and this bound consuming and producing countries together. This was coupled with massive investment overseas, all organized and backed up by an efficient and reliable financial system.

While all this was taking place world population increased from about 680 million in 1700 to 1.25 billion in 1850, to 2.5 billion in 1950, and to double that in 1987. Initially much of this growth was in Europe, but after 1945 most was in the less-developed world, that is, in the tropical world. Everywhere there was enormous pressure on the forests as a source of raw materials, and, when cleared, as land for agriculture. Again, the general conclusions are best given some reality by looking at individual case studies.

BRAZIL

The nibbling away at Brazil's coastal subtropical forest continued steadily throughout the eighteenth and early nineteenth centuries, but the principal causes of deforestation changed from gold-mining and sugar production. By 1830 coffee had passed sugar as Brazil's main export commodity, and by 1925 it accounted for two-thirds of the country's export earnings. The rapid expansion of cultivation and the attendant destruction of forests first hit the states of Minas Gerais and adjacent portions of Espirito Santo in the 1840s, the rapidity of expansion owing much to the widespread belief that the coffee bush needed the soils of newly cleared forest in order to grow successfully. But it was the construction of the railroads over the obstacle of the coastal ranges and into the interior of São Paulo in 1867 that had the greatest impact on the forests. As always, new lines of communication into the forests provided two-way traffic, each as destructive as the other. Not only did the railroads provide the means to get the coffee out to world markets; they also provided access to the new labor force of European migrants that flooded in after 1888, when slavery was abolished and there was a dearth of labor. It is possible that 140,000 square kilometers (54,000 square miles) of forest were cleared for coffee alone by 1931.

There is an understandable tendency to think that deforestation is a product of agricultural expansion, and that is basically true. But vast areas of forest were felled in Brazil, as elsewhere, for industrial purposes such as iron and steel smelting and fuel for railroads, and particularly for domestic fuel wood. It is difficult to be precise as to how much forest was destroyed for these purposes, but it cannot have been much less than that for agriculture. What we do know is that by 1970 only about 20,000 square kilometers (less than 8,000 square miles) of the primary stand of coastal subtropical forest remained out of an area originally nine times that size.

SOUTHERN ASIA

Despite their long-lasting independence from European influences, the countries of southern Asia could not escape the European impact entirely or forever. From the late eighteenth century onwards the British East India Company annexed more and more territory, and the princely states came under its economic rule. Gradually the economies of these areas became more commercialized and drawn into a wider world market, first with the cutting of teak for shipbuilding and construction, and then through the development of cash cropping with the institution of local taxes and the generation of land revenues. In contrast to the Spanish conquest of the Americas three centuries earlier, this was a relatively peaceful, benign, and stable administration in which both foreign and local entrepreneurs seized opportunities to make profits against a background of a rapidly increasing population.

It has been calculated that between 1880 and 1980 the population of the region (excluding southern India) rose from 190 million to 625 million. At the same time the forest area diminished from 888,000 to 542,000 square kilometers (343,000 to 210,000 square miles), and the amount of "interrupted" forest declined from 582,000 to 486,000 square kilometers (225,000 to 188,000 square miles). Most of the change was accounted for by the expansion of arable land, which rose from 757,000 to 1,253,000 square kilometers (292,000 to 484,000 square miles). With few exceptions, and unlike Brazil, the story of this loss is scarcely known because the expansion of arable land was incremental and gradual, and considered the most natural thing in the world.

One story, however, is both dramatic and documented. When the rainforests of the lower delta of the Irrawaddy in Burma came under British rule in 1852, government officials pursued a deliberate policy of clearing forest and mangrove in order to stimulate cultivation and provide rice for the adjacent famine-prone Indian mainland. Between 35,000 and 40,000 square kilometers (13,500 to 15,500 square miles) of tropical wet forest was systematically cleared, the land was drained and converted into rice paddies, and a new system of land tenure and taxation that was geared to encourage the local small landholders was instituted. The population of the region was 1.5 million in 1852 but increased to more than 4 million in the early 1900s, by which time the greatest bulk of the rainforest had disappeared.

Michael Friedel

AFRICA

So far nothing has been said about Africa. This is because the European impact here was relatively late

and probably never as great as in other tropical regions. Of course, the slave trade over centuries led to abandoned settlements, the depopulation of large areas, and possibly even to the regrowth and expansion of the tropical rainforest. But tropical Africa's humid, enervating climate, the prevalence of disease in humans and stock, and the enormous investment needed in getting the African territories into economic order discouraged European settlement, which either came late or was overshadowed by a "get-in-and-get-out" exploitive mentality as the continent was plundered for its minerals more than for its agricultural potential. Only in temperate southern Africa and the upland areas of East Africa (such as Kenya), where altitude ameliorated climatic extremes, was there any concerted European settlement.

Amerindian women in Brazil's Xingu River region carry shiny new metal pots, a product of Western technology, in traditional fashion on their heads. The real impact of Western technology on such forest dwellers remains to be seen, but experience around the world shows that the transition will not be easy.

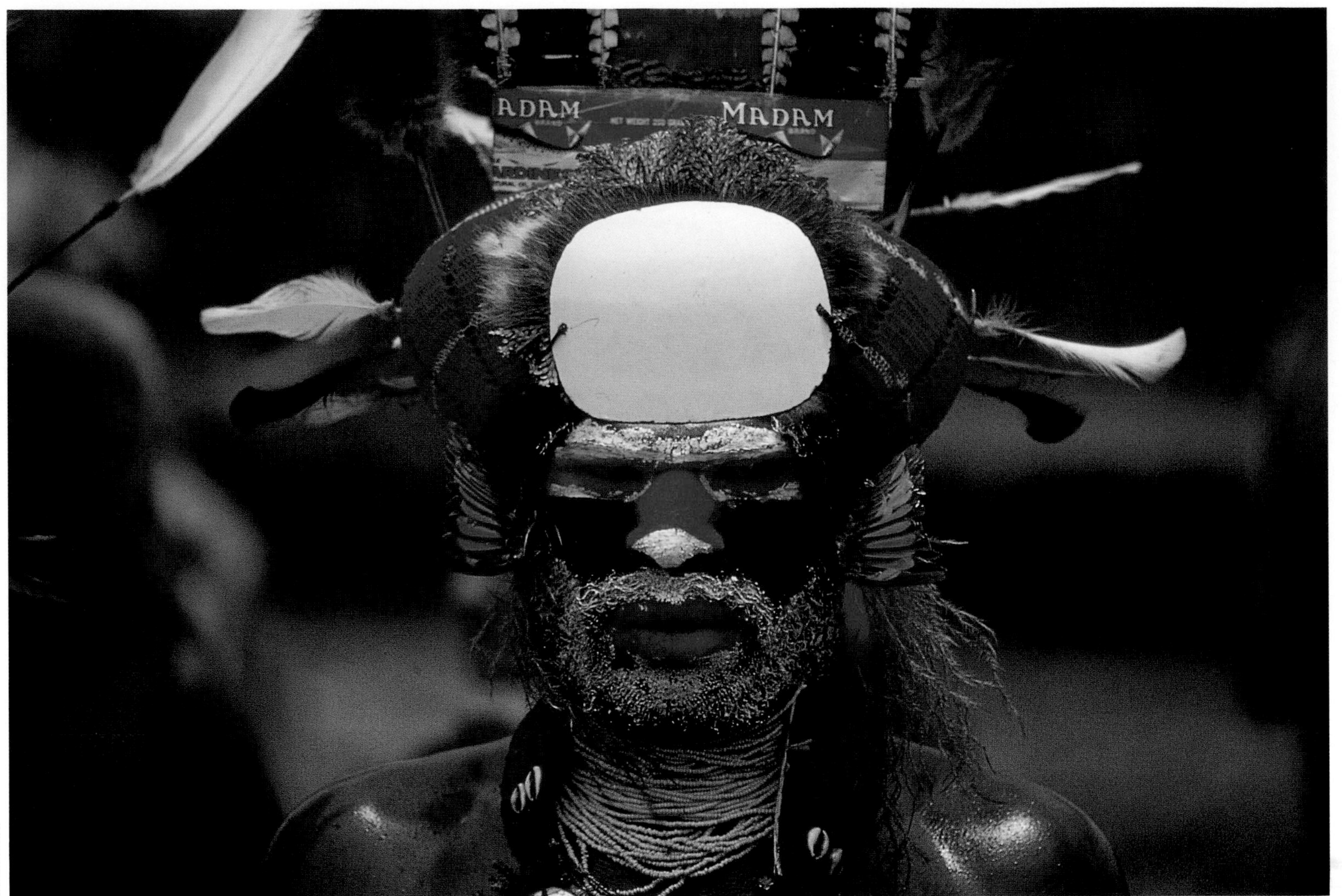

Michael Friedel

Above. *In the highlands of New Guinea, frequent celebrations,* sing-sings, *are an integral part of a traditional way of life that has existed for millennia. For these gatherings, men paint their faces and construct elaborate wigs and head-dresses. The impact of European contact is evident in the commercial dyes now frequently blended with traditional oxide pigments, and in the found objects, such as the canned sardine wrapper in this head-dress, often incorporated into the designs.*

Right. *The introduction of gold and diamond mining in Brazil meant an upsurge in the use of slave labor as depicted here in this eighteenth-century scene in Rio de Janeiro, where slaves wash gems under the watchful gaze of their colonial supervisors.*

Bibliothèque Nacional de Rio de Janeiro/Photographie Giraudon

The Last Sources of New Land

Concern at the European impact since the beginning of the voyages of discovery has given way during the latter half of this century to concern at the human impact, European or otherwise.

The reasons are clear. Foremost has been the startling reduction in mortality as a result of the introduction of Western medicine. This reduction has caused the population of the countries of the developing (largely tropical) world to more than double from 1.5 billion to 3.9 billion between 1945 and 1990. The press of humanity on the world's resources has become awesome. Consequently, pressures on the tropical forests have increased as they are perceived to be one of the last sources of new land for the extension of agriculture, for fuel wood for heating and cooking and even for industry, and as a source of hard currency from the export of logs and wood chip products. Nor are those pressures likely to abate in the immediate future. It is estimated that the world population will leap from the current 5 billion to over 10 billion by 2050, and that two-thirds of that increase will take place in the developing world.

Loren McIntyre

Slash-and-burn agriculture is not necessarily indiscriminate. The Kayapó of Brazil, for example, prepare their land according to a distinct timetable: the forest is usually cut in April or May, then left to dry until late September or October, when the land is fired. Some time later charred branches and debris are gathered up and set afire again to produce especially fertile "hot-spots". The entire process is carefully monitored by shamans who select the precise timing according to a range of biological indicators, such as the flowering of particular forest trees.

At the same time the people of Europe and other countries of the developed world have become concerned over the environmental consequences of development, growth, and the exploitation of resources. Many have become aware of the finite nature of the Earth's resources, this awareness being aided graphically by the advent of space photography and imagery that shows burning and clearing over vast areas. Concern at the loss of biodiversity has also grown. And there is a third element to the equation. Improvements in technology have increased the impacts on the forests. The chainsaw, the truck, and the tractor have enabled even the most inaccessible areas to be felled for whatever reason, and almost single handed. Advances in chemical technology and heat processing have enabled even poor quality wood fibers to be made into particle boards, laminated timbers, and cardboard. Thus, species of tropical trees that would never have been cut because of their inaccessibility, low yields, or poor quality timber are now worth exploiting. In short, all trees have a commercial value, all trees can be harvested, and the result has been an unprecedented onslaught by shifting cultivators, pastoralists, peasant agriculturalists, and loggers.

The extent to which that clearing is a result of European impacts is open to debate. If one wanted to be controversial one could argue that just as the initial incursion of Europeans into the New World reduced the population through the unintentional introduction of deadly diseases, so the incursions of European medical technology into the tropical world after 1945 reduced disease and increased the population. Both are true, but neither, it seems, gets us much further as each was irreversible. Secondly, one could argue that just as in the past, so at present, the capital and markets of the developed world provide the reason for much clearing. For example, the markets for cheap beef in fast foods and for durable furniture lead to pasture development in Central America and to hardwood forest felling in Southeast Asia.

By and large, however, the most potent cause for tropical rainforest destruction is the simple fight for survival by the inhabitants of the tropical world as they struggle for food, fuel, and cash. It is on those basic needs that a new European impact would be welcome, in the form of aid, conservation, and technology transfer. Then the tropical rainforests might have a chance of survival. ■

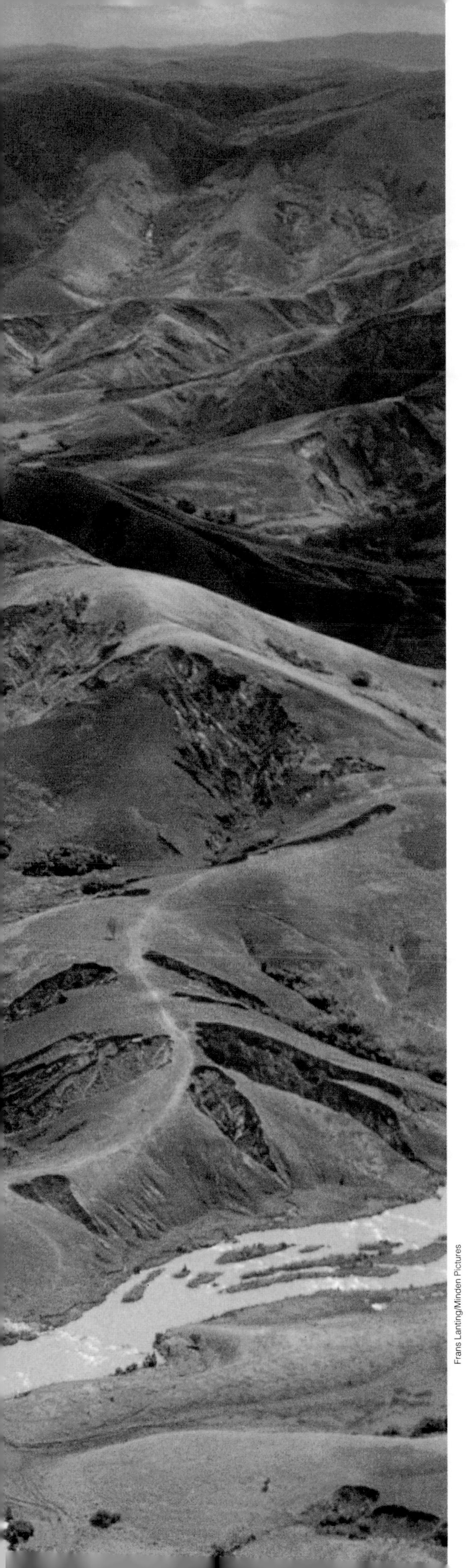

Frans Lanting/Minden Pictures

PART THREE

The Future of the Forests

The rainforests of the world have been reduced by more than half since the beginning of this century. In the past 10 years the rate of destruction has increased by 90 percent. The remainder of the forests are set to be eliminated within a few more decades, but there is a ray of hope. Never before has there been such worldwide interest in and enthusiasm for conserving the forests.

9 Nature's Greatest Heritage Under Threat

NORMAN MYERS

Tropical forests are disappearing. Not only that, they are disappearing faster than ever. According to a survey I conducted in late 1989 for Friends of the Earth (UK), the deforestation rate expanded by 90 percent during the 1980s. Worse still, the rate of increase is itself increasing. We have already lost half of the forests, and the rest are set to be largely eliminated within a few more decades.

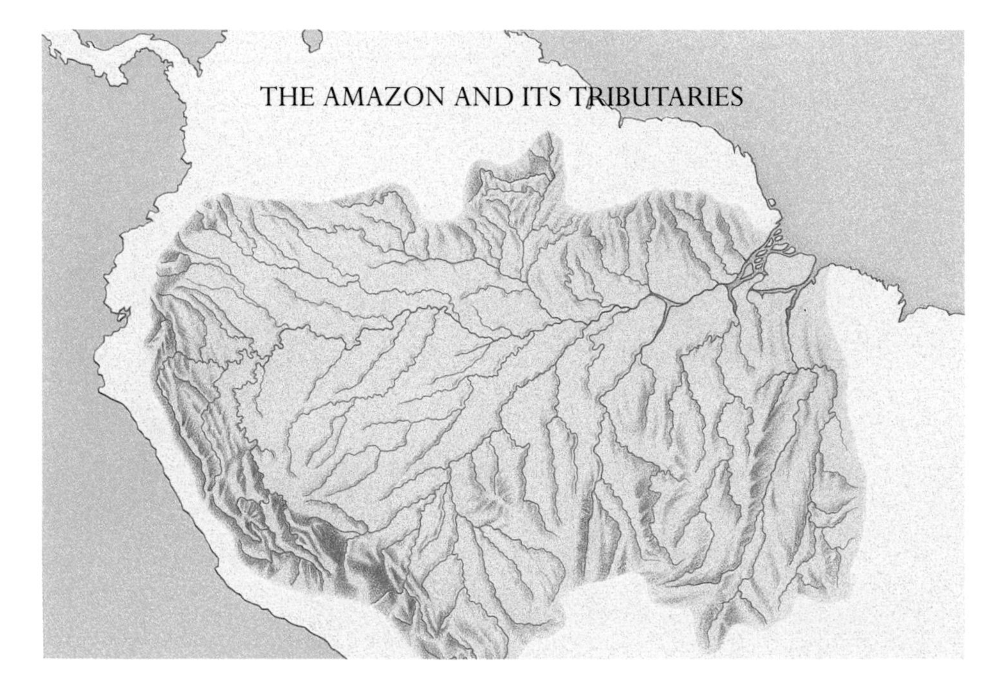

In terms of the total quantity of water transported, the Amazon is by far the largest of all rivers, discharging into the sea one-fifth of the combined total of all fresh water running into all the oceans of the world.

Opposite. *A rainforest in New Guinea being cleared for slash-and-burn agriculture. Recent studies indicate that during the past 10,000 years virtually all Asian forests have been affected by human activities such as slash and burn. Population pressures have greatly accelerated this activity in recent decades, but the practice in itself is not necessarily threatening the forest and its animal inhabitants. Large Asian animals, for example, are adapted to feeding in clearings, and therefore profit from such land use.*

The Embattled Forests

Think for a moment of that vivid band of green we used to see in our school atlases, denoting the tropical forest realm as the most exuberant expression of nature to adorn the planet in the 4 billion years since life began. Such maps may have to be recolored—the brilliant green replaced with dirty brown to indicate that what was once the glory of the Earth is no more. That is the dreadful news. The better news is that, even though the problem has been growing worse during the past decade, the period has also witnessed a sunburst of interest and enthusiasm on the part of both the public and political leaders the world over. Hundreds of citizen activist groups, with millions of members, are now proclaiming their support for the forests; and at world summit meetings, tropical deforestation is now on the agenda with inflation, international trade, and terrorism as a top-level concern. With commitment of that sort, there must still be hope, but the tropical forests are at great and increasing risk and we can only assess support for saving the forests after we have examined the extent of the threat.

We are now losing at least 142,000 square kilometers (55,000 square miles) of tropical forests a year. This is an area equivalent to half the British Isles. The rate of destruction translates into 27 hectares (67 acres) per minute, which means that during the half hour it takes to read this chapter and look at its pictures, we can expect to lose 810 hectares (2,000 acres) of forest. The annual deforestation also amounts to 1.8 percent of remaining forests, which cover just under 8 million square kilometers (3 million square miles) or an area roughly the same as that of Australia or the United States. But note that a rate of 1.8 percent does not mean we can wait another 56 years before witnessing the final elimination of all remaining forests. The annual deforestation rate is accelerating, and may be as high as 4 percent per year by the turn of the century. Even more important, the destruction pattern is far from even throughout the zone. Within another half dozen years there could well be little forest left in most of the Indian subcontinent, East and West Africa, and Central America. Within another 20 years we should anticipate there will be horizon-to-horizon deforestation in most of Southeast Asia extending from the Philippines to Burma; and huge swathes will have been lopped off the eastern and southern sectors of Amazonia in Brazil, and off western Amazonia in Bolivia, Peru, Ecuador, and Colombia. But in much of the island of New Guinea, the Zaïre Basin, and the western portion of Brazilian Amazonia plus the Guyana highlands, there may be few pressures to devastate the forests for several more decades, if only because of sparse human populations.

The overall situation is dominated by 10 countries that are losing forest at a rate of 4,000 square kilometers (1,500 square miles) or more each year. These are Brazil, Colombia, Mexico, Burma, India, Malaysia, Thailand, Indonesia, Nigeria, and Zaïre. Their collective total is more than 105,000 square kilometers (40,500 square miles) or 77 percent of tropical deforestation. Of this total, Brazil accounts for somewhere between one-quarter and one-third; but as Brazil possesses one-third of all remaining forests, this is not so surprising. In fact Brazil's annual deforestation rate of 2.3 percent is not as high as that in 13 other countries. The Philippines and Vietnam feature rates above 5 percent per year, Thailand and Madagascar

The diagram above shows the comparative loss of tropical and temperate forest cover over a 50-year period. By the year 2000, the projected coverage of tropical forest is a mere 7 percent. This contrasts sharply with the temperate-forest area, which covers, and in the year 2000 is still expected to cover, 20 percent of the globe.

above 8 percent, and Nigeria and the Ivory Coast above 14 percent. By contrast, 11 countries feature a rate of 1 percent or less: Peru, Venezuela, the Guianas, Cambodia, Papua New Guinea, Zaïre, Gabon, and the Congo.

Agents of Destruction

The best-known deforesters are the commercial loggers who exploit an additional 45,000 square kilometers (17,500 square miles) of forest each year. Some of this activity is light logging, taking only a few trees per hectare, causing little disruption, and allowing the forest ecosystem to recover fairly quickly. But in a good 30,000 square kilometers (11,500 square miles), mainly in the dipterocarp forests of Southeast Asia, commercial logging is so heavy that it takes out at least one-quarter of the trees, and the logging techniques are so disruptive that they leave two-thirds or even three-quarters of the remaining trees injured beyond recovery. The resultant tropical forest is a travesty of the name. It is a wreckage from which forest ecosystems cannot fully recover for decades, if not far longer. So the heavy logging is effectively destroying all those 30,000 square kilometers (11,500 square miles) of forest a year. Fortunately this logging-derived deforestation hardly increased in annual extent during the 1980s. The second agent of destruction is the cattle rancher of Central America and Amazonia. Ranchers in this area burn off roughly 15,000 square kilometers (about 6,000 square miles) of forest a year. Again, the amount hardly increased throughout the 1980s. Other small-scale deforesters include road builders, dam builders, and miners, who together account for about 12,000 square kilometers (4,500 square miles) of forest a year.

The remaining 85,000 square kilometers (33,000 square miles) of deforestation—60 percent of the total—are the work of the slash-and-burn farmer. The problem here is not so much the shifting cultivator, that is, the traditional farmer whose migratory lifestyle allows forests to recover after his transient impact. Rather, the problem lies with the "shifted" cultivator, the impoverished peasant who finds himself landless in his country's established farmlands, and has no option but to pick up his machete and matchbox and head for the last "free" and unoccupied areas.

This unfortunate farmer is driven by a number of forces that he understands little and can control even less. While part of his problem lies with the sheer buildup of human numbers, part too lies with the inequitable distribution of

The rainforest floor is protected from extremes of temperature, light, humidity, and wind by the vast bulk of foliage above. Clearing or other forms of destruction alter this stable regime, with far-reaching effects on the soil itself—an alteration that is in many cases irreversible. In the natural case, certain pioneer species establish themselves to gradually heal small-scale damage such as that caused by landslides or swidden agriculture. But when completely cleared for intensive farming, the soil may become totally impoverished within a few years, useless for anything but supporting low scrub.

low-grade scrub
total forest destruction
uncut forest

established farmlands. In Brazil, for example, 10 percent of farmers own 90 percent of farmlands, while the remaining 90 percent of farmers have to make do with only 10 percent of farmlands. Wealthy landholders may occupy thousands of hectares, finding it worth their while to cultivate only half their holdings, whereas the peasants over the fence try to support their families off just a few hectares.

The "push" factors driving the shifted cultivator into the forests are at the heart of the deforestation problem. If we were to eliminate the commercial logger, the cattle rancher, and the other minor actors, we would still have solved less than half of the problem. Nevertheless, these minor actors can all make their economic way by other methods: commercial loggers can grow timber in plantations on already deforested land, and cattle ranchers, through a modicum of management, can raise twice as many cattle on existing pasture land. No such ready option is open to the shifted cultivator.

Shifted-cultivator communities have a total population of at least 200 million people. Some observers estimate there could be as many as 500 million. The fact that we have no worthwhile estimate of their true total is a measure of how far their plight has been ignored. What we do know is that the migrating multitudes are expanding ever more rapidly. This is to be expected, if only because the world's population is growing so quickly. By the year 2030 or thereabouts, 80 percent of the world's projected population of 8 billion people are expected to be living in tropical-forest countries. This translates into 6.4 billion people, or many more than the total now living on Earth.

Everybody's Hand on the Chainsaw

It would be hardly realistic, and less than fair, for developed-world people to point a critical finger at the governments of tropical-forest countries, as if the entire responsibility for deforestation rested with them alone. Outsiders play their part as well, however unwittingly at times: hardly any of us does not at some time lend a hand to the chainsaw at work in tropical forests.

For one thing, people in developed countries stimulate overexploitation of tropical timber through their marketplace demand for specialist hardwoods from tropical forests. The reader might consider his or her own home, which may well feature a parquet floor, fine furniture, luxury paneling, and all manner of cabinets and other fittings, many of which contain tropical hardwoods. (There is even a good chance that the last raw material a citizen of the affluent world uses will derive from tropical forests: many coffins are made from imported hardwoods such as West African abachi wood and Brazilian mahogany.) Developed countries could grow more hardwood timber of their own, though it would be at a greater cost than tropical timber. In the meantime, let us consider that the price the developed-world citizen pays for tropical-forest hardwood does not reflect the full cost of production, especially the built-in cost of forests destroyed in Borneo, West Africa, and Amazonia.

In a more indirect fashion but with similar result, North Americans have been contributing to the loss of tropical forests through their desire for cheap supplies of beef, especially in the form of fast foods such as hamburgers and frankfurters. Since the early 1960s, beef has been one of the most inflationary items in the American weekly shopping basket. As a result the United States government authorized imports of so-called cheap beef from Central America, and that beef is raised on pasture lands established almost invariably through the clearing of tropical forests. By trying to trim a nickel off the price of a hamburger, the United States has contributed—unwittingly, but effectively and increasingly—to the massive loss of forests from Mexico to Panama.

Fortunately, there is a good-news aspect to this "hamburger connection". An environmental group in San Francisco, the Rainforest Action Network, mobilized a consumer boycott against the largest importer of Central American beef, Burger King. It took two years of persistent effort on the part of millions of individual Americans before

forest cut and burned — *farm in use (2–3 years)* — *two years later, pioneers established* — *after 15 years, small primaries emerge* — *after 60 years, primaries dominate* — *after 100 years, similar to uncut forest*

Steve Cox/Camera Press London/Austral

A Kayapó Indian addresses a rally in Amazonia. Many such rallies are now held worldwide as indigenous peoples mobilize in defense of the forests they have inhabited for millennia. The portrait at lower right commemorates Chico Mendes, the Brazilian who fought to save the Amazon forests for rubber tappers and whose assassination attracted worldwide media attention.

success came their way, and during that time many of them must have had occasion to ask themselves whether they would finally win through. But they stuck at their task, determined to register their conservationist vote in the marketplace. Eventually Burger King gave in, vowing to import no more of the artificially cheap beef. This exceptional success story should hearten us whenever we suspect the deforestation problem is so big that it will surely withstand the one-by-one actions of citizens. Remember the remark of Edmund Burke: "Nobody ever made a greater mistake than he who did nothing because he could only do a little."

The hamburger connection in the Americas is matched by a similar one between the European Community and Southeast Asia. Europeans, with their desire for cheap supplies of livestock feed, have been importing millions of tonnes of calorie-rich cassava from Thailand every year, to be fed to their excessively large stocks of cattle, pigs, and poultry. The cassava is grown in Thai croplands which have been established mainly at the expense of forests. Again, the cassava-fed beef, pork, and chicken amount to a "free lunch" in some respects, as the price does not cover all production costs.

These are some of the connections between the way people live in developed countries and the way tropical forests are fading away. Each of us needs to learn more about the "economic ecosystems" of the international marketplace. It is difficult for us to see these relationships in principle, let alone track them down in practice, but until we can achieve "systems thinking" along these lines, our consumerist left hands will continue to undo much of the good accomplished by our conservationist right hands. It is right and proper that we should shed a tear over the plight of tropical forests; but while shedding that tear, let's go easy on the Kleenex.

We might, for instance, consider our linkages to the banking community. International debt on the part of tropical-forest countries totalled more than US$550 billion in 1990. So great is this millstone around the neck of their economies that their interest payments to the rich countries exceed the foreign aid they receive by more than US$50 billion a year. The longer such a situation persists, the less are the chances that tropical-forest governments will be able to divert funding to subsistence agriculture in their countries and thus help the plight of the impoverished peasant who is poised to become a shifted cultivator.

What is at Stake?

Just as we are all involved in tropical deforestation, so we shall all lose if the forests continue to disappear. The main adverse repercussion at a global level will likely emerge in the form of climatic dislocations. But there are several other potentially disastrous results.

The first concerns hardwood exports from tropical-forest countries. The decline of the forests has already caused several dozen countries to lose a major source of foreign exchange. Tropical hardwood exports earned US$8 billion in 1980, making this timber as valuable as cotton, twice as valuable as rubber, and almost three times as valuable as cocoa exports. At the time of writing, in the wake of deforestation, these export earnings have slumped to only US$6 billion, and they are likely to decline to US$2 billion by the year 2000, after which time tropical-forest countries may become importers rather than exporters of timber. For those many countries where hardwood revenues play a key role in economic advancement, the impact of deforestation will be devastating.

Next, deforestation leads to a decline of environmental services, notably watershed functions. In the Ganges River system, deforestation of river catchments in the Himalayan foothills contributes to flood-and-drought regimes for 500 million small-scale farmers in India and Bangladesh, with costs in India alone estimated at more than US$1 billion a year. Several other river systems of tropical Asia, notably the Salween and Irrawaddy in Burma, the Chao Phraya in Thailand, and the Mekong in Vietnam, have similar deforestation-related problems, although on a smaller scale than in the Ganges valley.

Lorer McIntyre

As currently practiced in most tropical areas, logging is unsightly and wasteful, although what remains after the loggers have left is not necessarily a biological desert. Some logged forest remains able to support at least some of its original plant and animal inhabitants, and much room remains for the development of techniques for reducing the destructive impact of logging and stimulating regeneration.

Decline of environmental services is also apparent in a number of non-agricultural development sectors. For example, increased sedimentation as a result of deforestation-induced erosion of topsoil reduces the operational life of giant dams for hydropower. Instances in the Philippines, Thailand, India, Pakistan, Kenya, Colombia, Ecuador, and Central America show that the collective costs run into billions of dollars a year. Similarly, washed-off silt and other soil debris exert a "smother effect" on coastal fisheries. Deforestation can also endanger public health in areas where water-related diseases are rife and tropical forests are the main or only sources of good-quality water for domestic use.

But the longest-lasting repercussions will stem from the mass extinction of species—an event that has already begun in the tropical forests. Non-economic factors apart, the extinction of species means the irreversible loss of uniquely diverse raw materials in the form of those species' genetic resources. Madagascar is home to the rosy periwinkle, the properties of which led to the development of two anti-cancer drugs that have commercial sales of US$170 million per year in the United States alone. But deforestation has already eliminated 90 percent of Madagascar's forests, and the island has almost certainly lost or is about to lose 2,500 plant species found nowhere else, among them plants with fertility-control properties and other high-value attributes. Cancer specialists believe there are at least another 20 plants in tropical forests with therapeutic capacities against various forms of cancer. When India's Silent Valley forest was saved through the initiative of citizen activists, it was subsequently found to harbor a wild relative of rice with genes that helped combat a disease that threatened Asia's rice crop.

The work of pharmacognocists (those who study drugs of plant and animal origin) is still in its infancy—only one plant species in 100 has been assessed for possible applications. Already, hundreds of genetic materials that can contribute to industry and bioenergy as well as to medicine and agriculture have been isolated. However, when a species has gone—and all the evidence suggests many have become extinct even before they have been isolated—it has gone for good.

On top of these materialist considerations, there is a moral objection to species extinctions. Deforestation is well on its way to triggering a mass extinction to match if not surpass the greatest mass-extinction episodes in the prehistoric past. Of course evolution and natural selection will eventually come up with a replacement stock of species, but so far as we can discern from the

WILL THERE BE A MASS EXTINCTION IN THE FOREST?

No other biome, or ecological zone, is as rich in species as that of the tropical forests. While they occupy only a small percentage of Earth's land surface, the forests contain at least 70 percent and perhaps even 90 percent of all Earth's species. At the same time, no other biome is undergoing such rapid destruction. A mass extinction is taking place, and it may well eliminate many more species than are found in the rest of the world.

KEY TO ANIMALS

1 aye-aye *Daubentonia madagascariensis*
2 golden toad *Bufo periglenes*
3 orang-utan *Pongo pygmaeus*
4 mountain gorilla *Gorilla gorilla beringei*
5 maned sloth *Bradypus torquatus*
6 woolly spider monkey *Brachyteles arachnoides*
7 golden lion-tamarin *Leontopithecus rosalia*
8 monkey-eating eagle *Pithecophaga jefferyi*
9 Bannerman's turaco *Turacus bannermani*
10 Malayan tapir *Tapirus indicus*
11 tiger *Panthera tigris*
12 indri *Indri indri*
13 Queen Alexandra's birdwing butterfly *Ornithoptera alexandrae*

There are 14 areas in the world that feature exceptional numbers of species found nowhere else, and that face imminent deforestation. These areas include western Ecuador, the Colombian Choco, western Amazonia, Atlantic-coast Brazil, southwestern Ivory Coast, the montane forests of Tanzania, Madagascar, the Western Ghats mountains of India, southwestern Sri Lanka, the eastern Himalayas, Peninsular Malaysia, northwestern Borneo, New Caledonia, and the Philippines. The entire expanse of these areas comprises 311,000 square kilometers (120,000 square miles), or less than 4 percent of remaining tropical forests. Yet they contain almost 37,500 endemic plant species, or 29 percent of all the 130,000 plant species found in tropical forests. Generally speaking, we can reckon there are at least 20 animal species, and more likely 50 or even 100 (mostly insects), for each plant species. And the same ratio applies to endemism. Taking the most conservative figure, this means these areas feature 750,000 animal species found nowhere else. All in all, then, the deforestation of these 14 "hot-spot" areas would precipitate a species extinction spasm on a scale unmatched since the "great dying" of the dinosaurs 65 million years ago. Already we are losing one plant species and 20 (minimum estimate) animal species in these areas every day, or well over 7,500 species per year.

Above. *Skulls of extinct Madagascan lemurs.*

Salomon Cytrynowicz/D. Donne Bryant/Stock Photos

Waving the Brazilian flag as a banner, a deputation of indigenous Amerindians gathers at a conference in Brasilia. The growing political sophistication of many Amerindian tribes introduces a potent new factor into the global struggle to halt the destruction of the rainforests.

geological record, the time needed for evolution to do so would be at least 5 million years, possibly twice that.

In other words we are engaged in an impoverishment of our biosphere that will not be made good within a period at least ten times longer than humans have been humans—and we are doing it in less than a single century. Is this not a factor that should give us pause?

What Will Save the Forests?

At least 60 percent of present deforestation is due to the shifted cultivator. Moreover, the proportion appears set to rise rapidly. The overwhelming factor driving these displaced peasants into the forests and into destroying them is land hunger. While this reflects population growth, it also stems from the unfair distribution of established farmlands. So a key to saving the tropical forests must lie with agrarian reform. Also helpful would be increased support for subsistence agriculture to help the small-scale peasant make more intensive use of his existing farm plot, small as it is. Further, he needs better marketing networks, credit facilities, extension services and a host of other rural-development measures that would relieve him of his motivation to head for the forests. It is measures such as these, rather than more parks and reserves, that will ultimately save the forests.

In short, the phenomenon of shifted cultivators represents a failure of development strategies across a broad spectrum. In the long haul, their problem can be confronted only by a major restructuring of development policies and planning. It may sound strange for would-be saviors of tropical forests to urge the cause of land re-distribution in territories many horizons away from the forests. But since the forests are falling mainly for non-forestry reasons, the question of their survival is no longer a matter to be answered by foresters or conventional conservationists alone.

The problem of deforestation worsened during the 1980s, and prospects are bleak. But an extraordinary outburst of concern and support for reform on the part of people right around the world makes it appear that at long last we may be ready to get on top of the problem before it gets on top of us. First of all, there has been one series after another of citizen activities by on-the-spot groups in tropical-forest countries. In India the Chipko people or "tree huggers" have beaten back the loggers who planned to fell some local forests in disregard of local needs. The community in question valued the forests for their fruits, fodder, medicinal materials, and fuelwood, all of which could be harvested sustainably, and they saw little benefit for themselves in watching the trees being converted into lumber for remote commercial interests. It took several years to persuade the loggers to look elsewhere but finally the Chipko people, most of them semi-literate peasants, won the day. Their success was subsequently replicated by

Michael Friedel

another grassroots initiative at Silent Valley in southern India, when local people halted a dam that would have flooded a large part of their forest.

In Kenya the Greenbelt Movement, run entirely by women, planted more trees in its first year of operation than the government had during the previous 10 years. This exercise relieves excessive exploitation pressure on remaining forests. There is a similar Greenbelt Movement in Colombia, and it has enjoyed equal success. In Indonesia there are 400 local conservation groups, which collectively have sufficient clout at national level to regularly gain the ear of government ministers. In Costa Rica, Ecuador, and a lengthy list of other countries there is a similar burgeoning of grassroots activism. Even in Brazil there is a flourishing community of non-governmental bodies which the government feels it must heed.

Kuikuru Indians from the Xingu River region in Amazonia slash and burn a clearing for farming. This traditional method of farming is often named as a major cause of tropical deforestation, yet it has been practiced for centuries, which is evidence that it is a sustainable system. The indigenous people clear only small portions of land in isolated patches of forest. The areas are intensively farmed for up to 30 years; the plot is then reabsorbed into the forest and a new site is prepared.

Citizen supporters of tropical forests are on the march in the United States. In 1985 the Rainforest Action Network was almost alone in its campaigning. Today there are more than 100 rainforest action groups around the country, mainly on college campuses, and another springs up every week or so. Parallel stories can be told in Britain, Holland, Germany, Sweden, and a host of other non-tropical countries where there is increasing recognition of there being a common responsibility for the common heritage in those far-off forests.

As for political leaders—who are often so ecologically illiterate that they think a food chain is a line of supermarkets—they now sense they have no option but to follow the spirit of the times. The heads of governments in the Philippines, Thailand, India, and Kenya have declared that deforestation constitutes a "national emergency". The Colombian government has handed over half of its Amazonian forest to the care of tribal peoples in the light of their demonstrated capacity to make a living out of the forests without knocking them over. Among the developed nations, the British government has assigned an extra £100 million to tropical forests; the German government is mounting a large-scale program to support the cause; and in Holland, Sweden, Norway, Italy, and Canada there are signs that governments are ready to play a much more solid part in the global campaign. So, even as more forests than ever are dispatched in smoke, there is a worldwide movement to confront the challenge. We must not overplay the new initiatives. They are no more than a start. But things look far more promising than they did in the dismal days of the early 1980s, when tropical forests were on the agenda of nobody but the loggers and their like. ■

10 Safeguarding our Heritage

JEFFREY A. MCNEELY

The pace of destruction of tropical rainforests is accelerating, and political pressure at local, national, and international levels has finally made it imperative for governments to act in their defense. The question is no longer whether to conserve tropical forests, but rather how it can be done, how much it will cost, and who will pay the bill.

WWF/Mark Edwards

This Forest Protection Group poster represents the growing ranks of rubber tappers, Indians, fishermen and other forest workers of Amazonia who are rallying to defend the forests upon which their livelihoods depend.

A Serious Business

Logging, shifting cultivation, conversion to agriculture or pastures, and weak forestry departments are usually seen as the main threats to tropical forests. But these are just manifestations of the far more fundamental problem of how the resources of the tropical forests should be allocated.

Successful conservation of the tropical forest must involve a "new forestry" that assesses and balances the needs of the currently antagonistic groups. Should the benefits from using the forest go primarily to the timber concessionaires, large plantation owners, and government forest departments, with some benefits trickling down to local people in the form of schools, roads, and health care? Or should most of the benefits go directly to the people who both preserve and live in the forest, with additional benefits being passed on to the nation in the form of taxes, income from tourism, improved water supply, and trade goods? Or is some balance between these two options possible?

During recent decades, most governments have sided with the interests of rapid forest exploitation, using subsidies and other economic incentives to accelerate the process and earn quick profits. The interests of the local people have received relatively little attention. But several studies have now shown that the "quick buck" approach is economically flawed, because the full costs of forest destruction have not been considered. Much of the profit has come from subsidies for clearing land, or from insufficient concession fees that have not covered the costs of repairing the environmental damage done by logging: disruption of water cycles, effects on local and global climate, and damage to wildlife populations. Investments were not made in replanting the forest, improved research, or management, protection, and compensation for forest peoples.

But, governments are finally beginning to listen to the local people who see their forest disappearing before their eyes, to the economists who argue for a more complete assessment of tropical forest values and the costs of overexploitation, to the climatologists who are worried about the impact of deforestation on the global climate, and to the conservationists who argue for the survival of the planet's species and ecosystems.

Conserving What's Left

It is almost too late for those countries whose forests have been reduced to mere remnants. In Sri Lanka, Thailand, Sierra Leone, El Salvador, and many other tropical countries, forests have been so overexploited that conserving the remainder is the only option. Other countries, such as Brunei, Indonesia, Zaïre, Gabon, Brazil, Colombia, and Surinam, still have considerable forests remaining and more options for the future. But even in these countries, conserving tropical forests is no easy matter. Simply establishing protected areas or banning the trade in tropical timber will not conserve the forests. Important as they are, national parks and other protected areas cannot be expected to prosper unless the underlying social, economic, and political causes of the loss of the forests are addressed. Otherwise, a halt to logging may only result in forests being put to other uses that may be even more damaging.

Conserving tropical forests requires a fundamental change in priorities. Given the importance of tropical forests to people, the growing public support for conserving them, and the very heavy pressures that are being put upon them, it is clear that the time is right for moving to a new relationship with tropical forests that encompasses defending, understanding, and using them wisely and equitably. Such an approach must include: the establishment of legally protected areas to conserve the tropical forests that are particularly important for their biological diversity; the utilization of tropical forests in a sustainable way; mechanisms to allow the rural communities that are most dependent on the forests to manage them for a sustainable yield; and the education of all people about forest functions and the consequences if they are destroyed.

This new forestry is being supported by international action, in the form of legislation, technical assistance, and financial support. And while it may not appear as dramatic as battles with poachers or marches against government policies, it is far more likely to lead to successful and sustainable forest management. As the exploitation of tropical forests has accelerated, governments have come to realize that the establishment of protected areas of relatively intact forest is a necessary part of balanced land use. Most often, these areas take

Opposite. *Planting Caribbean pine seedlings in the ashes of a Brazilian rainforest. Modern plantation techniques have the potential to produce pulp and industrial roundwood far more efficiently than by clear-felling virgin rainforest. It has been estimated that the world's total demand for pulp could be met from exploiting about 5 percent of tropical forest lands.*

CATEGORIES OF PROTECTED AREAS

While all protected areas control human occupancy or use of resources to some extent, considerable latitude is available in the degree of such control. The following categories are arranged in ascending order of degree of human use permitted in the area.

CATEGORY I: SCIENTIFIC RESERVE/STRICT NATURE RESERVE
To protect nature and maintain natural processes in an undisturbed state in order to have ecologically representative examples of the natural environment available for scientific study, environmental monitoring and education, and for the maintenance of genetic resources.

CATEGORY II: NATIONAL PARK
To protect natural and scenic areas of national or international; significance for scientific, educational and recreational use. These are relatively large natural areas not materially altered by human activity and where commercial extractive uses are not permitted.

CATEGORY III: NATURAL MONUMENT/NATURAL LANDMARK
To conserve nationally significant natural features because of their special interest or unique characteristics. These are relatively small areas.

CATEGORY IV: MANAGED NATURE RESERVE/WILDLIFE SANCTUARY
To maintain the natural conditions necessary for the existence of nationally significant species, groups of species, biotic communities, or physical features of the environment when specific human manipulation is required for their perpetuation. Controlled harvesting of some resources may be permitted.

CATEGORY V: PROTECTED LANDSCAPE
To maintain nationally significant natural landscapes characteristic of the harmonious interaction of people and land, while providing opportunities for public enjoyment through recreation and tourism within the normal lifestyle and economic activity of these areas.

CATEGORY VI: RESOURCE RESERVE
To protect the natural resources of the area for future use and prevent or contain development activities that could affect the resource, pending the establishment of objectives based on appropriate knowledge and planning.

CATEGORY VII: NATURAL BIOTIC AREA/ANTHROPOLOGICAL RESERVE
To allow the way of life of societies living in harmony with the environment to continue undisturbed by modern technology: resource extraction by indigenous people is conducted in a traditional manner.

CATEGORY VIII: MULTIPLE-USE MANAGEMENT AREA/MANAGED RESOURCE AREA
To provide for the sustained production of water, timber, wildlife, pasture, and outdoor recreation, with the conservation of nature oriented to support economic activities (although zones can be designed within these areas to achieve specific conservation objectives).

Source: The World Conservation Union (IUCN)

PROTECTED AREAS IN TROPICAL FORESTS

Categories are assigned following IUCN criteria, in ascending order of human activity. (No reliable figures are available for Categories VI, VII, and VIII.)

CATEGORY		NUMBER OF AREAS	AREA PROTECTED (HECTARES/ACRES)
Category I	Nature Reserves	125	10,469,546/25,304,893
Category II	National Parks	216	35,667,429/86,208,176
Category III	Natural Monuments	6	21,050/50,888
Category IV	Wildlife Sanctuaries	263	16,817,843/40,648,727
Category V	Protected Landscapes	59	3,067,537/7,414,237
	Total	669	66,043,405/163,127,209

Source: World Conservation Monitoring Centre, December 1990

the form of national parks. Indeed, such parks are an indispensable element of development because they maintain essential ecological processes—including soil formation, nutrient cycling, and cycling of water. They also preserve the diversity of species within environments where they can continue to evolve naturally.

But other kinds of protected areas also contribute to forest conservation. To illustrate the range of land management methods that may help maintain biological diversity while contributing to sustainable development, the World Conservation Union (IUCN) has developed eight categories of protected areas, each of which is designed to achieve an array of objectives.

In the two decades up to the end of 1990, 669 protected areas covering more than 66 million hectares (163 million acres) had been established in tropical forests. But very few new areas have been set up since 1985, which suggests that options for such areas are rapidly diminishing. If the current extent of tropical forests is taken as 8 million square kilometers (3 million square miles), this means that about 8.3 percent of tropical forests are currently protected to some degree. The general consensus is that this figure needs to be increased to at least 10 percent.

The importance of establishing protected areas in tropical forests has received increasing public support as their value and potential have become better understood. Intense lobbying by non-governmental organizations has halted some large development projects that would have destroyed tracts of important tropical forest and the human cultures found there. For example, a scheme to dam Silent Valley, considered to be one of the last representative stands of mature tropical evergreen forest in India, became the focus of one of the fiercest and most widely publicized environmental debates of the region. The project was finally shelved in 1983 in deference to the sentiments of the then prime minister, Mrs Indira Gandhi, and the site has since been designated a national park.

In some cases, development projects actually benefit or promote the creation of protected areas. In northern Sulawesi, Indonesia, the Dumoga-Bone National Park was established with funding from the World Bank to protect the watershed of an irrigation project downstream; and the Mahaweli Environment Project in Sri Lanka included a US$6 million component to establish a new system of national parks as part of the water resources development effort.

However, the simple proclamation of a national park is not enough. Because virtually all tropical forests are inhabited by humans—either traditional (tribal) groups or landless migrants from the nation's dominant culture—establishing protected areas also involves costs, particularly for those people living in the vicinity of protected areas who may be prevented from exploiting resources as freely as they might wish. Also, such areas may well be perceived as harboring "pests"—predators such as jaguars or tigers, which kill domestic livestock, or large herbivores, including elephants, which damage crops. Retaining natural vegetation also denies immediate benefits from logging and conversion to other forms of land use, though these losses may be less significant economically and ecologically, in the long term.

Still, if protected areas are to continue making their necessary contribution to human welfare, the distribution of costs and benefits (of both exploitation and conservation) needs to be more equitable. While the specifics will vary from case to case, local support for protected areas must be increased through such measures as revenue sharing, participation in decisions, complementary development schemes adjacent to the restricted region, and access to renewable resources.

The best approach then to the long-term preservation of these vitally important areas is to design and manage a range of protected areas, so that conservation aims can be supported within the overall fabric of social and economic development. Advances in conservation management during the coming decades will be primarily in the establishment, implementation, and improved management of protected areas where some human use is tolerated and even encouraged, or in new types of protected areas where degraded landscapes have been restored to productive use. Strictly protected regions (IUCN categories I and II) are unlikely to ever account for more than about 10 percent of tropical forests. But since permanent agriculture seldom occupies more than about a quarter of any nation's area, ample land exists for forestry, shifting cultivation, grazing, and other uses. When land set aside for such use surrounds national parks it can also help conserve biological diversity by acting as a buffer against more negative human influences.

Adam Woolfitt/Susan Griggs Agency

Kuna women embroidering in Panama. Unlike many other forest peoples, the Kuna of central and southern Panama had sophisticated social, political, and economic structures at the time of first contact with European explorers in the sixteenth century. Some of their lands are now incorporated in the Kuna Indian Park, set aside for their exclusive use.

The Bigger Picture

Few protected areas are self-contained, and recent advances in conservation biology have shown that, in isolation, they will not be able to conserve all or even most of their species diversity, genetic resources, and ecological processes. For example, because many vertebrates need large areas to support a viable breeding population, far greater expanses are required for conservation than modern societies are willing to remove from direct production.

Consequently, new approaches to linking protected areas to surrounding lands—including those used for timber production—are required. Management of a protected area and that of the land adjacent to it must be planned in tandem. The establishment of "transition zones", which regulate human activity in adjacent lands, is crucial for the conservation of biological diversity within the strictly protected core area.

This new forestry will typically involve protected areas becoming parts of larger regional schemes to ensure biological and social sustainability, and to deliver appropriate benefits to the rural population. Conservation management will seek a balance of benefits from logged, farmed, and protected land, and involve rural development, economics, land tenure, tourism, protected areas, tribal lands, and a range of other interests. The hoped-for result is a formula that will both benefit the nation's economy and allow a sustainable harvest from the forest.

TROPICAL FOREST PRODUCTS

JAMES A. DUKE

The world's tropical forests contain a wealth of known raw materials which could be rationally exploited through agroforestry or extractivism, yielding employment, income, and cultural survival to forest residents while preserving the forest. Agroforestry is a farming system whereby through a certain amount of management various forest products and sometimes conventional crops are harvested from the forest. Extractivism is the renewable harvesting of economically useful products from natural ecosystems. Both strategies depend on sustainable exploitation of the forest's resources. The concept of extractivism is not new. What is new is the idea of extractivism as a potential savior of the rainforest.

Loren McIntyre

Brazil nut trees are protected by law and are not felled when land is cleared. Curiously, the trees do not burn, and remain dotted about the landscape even after extensive burning.

• *Palm hearts* Peach palm *Bactris gasipaes* is a prime example of a forest species whose fruits and palm hearts, long utilized by Amerindians, are now finding their way out of the forest onto urban dinner tables. This palm can furnish fruits renewably year after year and, under ratooning (propagation by cutting back old growth), palm hearts for more than one season. In Costa Rica the hearts are canned commercially, and other types of palm are used in the growing palm-hearts industry. As well, there is at least one wild palm, the mil pesos *Jessenia bataua* of Panama and Colombia, whose oil is not only nutritionally similar to olive oil but is also half as expensive to produce.

• *Palm oils* are possible alternatives to petroleum and are being considered in certain tropical Third World countries. Optimists have suggested that 2 billion hectares (5 billion acres) of tropical land with a renewable yield of 25 barrels of palm oil per hectare (10 barrels per acre) could satisfy the world's fuel needs. Babassu palm, from tropical America, and nypa, from tropical Asian brackish swamps, have also been suggested as fuel sources when the petroleum runs out. At present less US fuel imports are derived from tropical forests. However, as long as energy conservation is maintained, oils and alcohols developed renewably from tropical agroforestry biomass farms could meet all US import needs.

Alain Compost/Bruce Coleman Limited

Above right. *Cacao pods in East Java. Cocoa originated in South America, but is now successfully grown in other areas of the tropics.*

Tony Morrison/South American Pictures

These oil-producing palm fruits are growing along the banks of the Rio Negro, Amazonia.

• *Nuts* Tropical American forests are still the sole source of brazil nuts and the third-largest source of cashews. Not only are these renewable resources of great economic value (their annual sales to the United States alone approaching US$300 million), but prudent harvesting interferes little with the lifestyles of forest peoples and is indeed crucial to the survival of the crop. Few, if any, nut species are as dependent on the survival of the rainforests as the brazil nut, and all attempts to grow them on plantations have failed miserably.

• *Beverages* containing caffeine could all be produced in the tropics. Chocolate, guarana, and maté are native to tropical America, coffee native to Ethiopia and neighboring areas, cola to Africa, and tea to Asia. Though few of these are now harvested from the wild, all could be grown in soil-preserving agroforests, with maté and guarana perhaps harvested renewably from extractive reserves.

• *Food colorings* If patents are any indication, then interest in natural colors and dyes has doubled since the mid-1970s. The renewable dye annatto (from tropical American *Bixa orellana*) is used to color cheese, bakery products, soups, sauces, pickles, smoked fish, and other foods. Currently, about 10 percent of the market for natural food colorants is supplied by tropical forest products, but agroforest products could satisfy the entire market.

• *Medicinal drugs* Fewer than 10 percent (rather than a possible 25 percent) of modern medicines stem from tropical rainforest species. A policy of extractivism or agroforestry in rainforest zones could see that figure raised since drugs such as atropine, berberine, diosgenin, dopa, emodin, podophyllotoxin, sanguinarine, scopolamine, and xanthotoxin are currently derived from temperate plant species but could be derived from tropical ones. In fact, more than 75 percent of natural-product medicines could be extracted from tropical forests, and many *could* replace synthetic drugs.

• *Pesticides* Tropical forests are full of natural pesticides, so why do synthetics account for almost the entire market? The answer is that synthetics are cheaper to produce at present, and they are more easily patented. But food processors, adept at removing nutrients from foods, could just as easily extract naturally occurring pesticides from the food chain. Currently we use pyrethrins (from *Chrysanthemum* grown at high altitudes and low latitudes), rotenones (from various tropical legumes), neem (from the tropical neem tree), and ryania (from *Ryania*). At the time of writing, tropical natural pesticides would not account for more than 1 percent of the world's pesticides, but 100 percent could be extracted from plant species grown in tropical agroforests and extractive reserves.

• *Sweeteners* Each American consumes the equivalent of 68 kilograms (150 pounds) of sugar a year, with non-calorific sweeteners accounting for almost one-fifth of that intake. Saccharin consumption is about 3,000 tonnes a year, and between 1984 and 1990 the share of the soft drink market occupied by diet drinks rose from 25 to 50 percent. Consumers are certainly interested in *natural* non-nutritive sweeteners derived from forest products such as *Lippia* and *Stevia* from America, *Momordica* from China, and *Dioscoreophyllum*, *Synsepalum*, and *Thaumatococcus* from Africa, and all these plants could be grown and harvested in extractive reserves. The legal sweetener business is worth billions of dollars a year in the United States alone, but as none of these natural sweeteners has been approved by the appropriate US authority, none is produced in this cost-effective and ecologically supportive manner.

• *Vegetable ivory* Could the tropical vegetable ivory (*Phytelephas* species) do as much to save the elephant as the tropical jojoba (*Simmondsia*) is doing for the sperm whale? With the continued support of agencies such as Conservation International the vegetable ivory industry could indeed get a much needed boost, having long been in decline. The industry was at its peak in 1929, with Ecuador alone exporting more than 2,000 tonnes. By 1941 only 500 tonnes were exported, the New York price having dropped to only 10 cents a kilogram. Increasing support has meant that vegetable ivory has reappeared in the marketplace. Buttons made of it cost 25 percent more than plastic ones, but they have increased appeal for "green" consumers. One factory in Ecuador, which was founded in 1926 and is now picking up after a long slump, produces about 2,270 kilograms (5,000 pounds) of vegetable ivory buttons per month, exporting the entire output, mainly to Japan, Germany, and Italy.

As with so many palms, the ivory palm has many uses. There is interest in it as a source of abrasives, and as a source of important phytochemicals. Incredibly, the ivory, before it hardens, is custard-like, and quite tasty, and the leaves can be used for thatch. •

Loren McIntyre

A boy hawks cashew fruit in an urban street in Brazil.

Below. *Vegetable ivory nuts are rock-hard and can be carved in much the same way as ivory, which is now banned from international trade.*

WWF/Bios/J.L. Ziegler

WWF/Elaine/Pierre Dubois

A woman picks pyrethrum daisies in Rwanda. Pyrethrum is one of the few natural pesticides currently used even though tropical forests are full of natural pesticides, and could provide 100 percent of the world's pesticide needs. At the moment natural pesticides account for a mere 1 percent of the market.

LOGGING AND WILDLIFE CONSERVATION: IS THERE ROOM FOR BOTH?

A recent study has estimated that less than 0.1 percent of all timber production operations in the tropics are sustainable. Even so, it is the short-sighted view of a number of governments under pressure for foreign-exchange earnings that timber is their forests' major economic resource. This does not necessarily mean that logged areas are lost to conservation. Selective logging need not destroy all native species, or even most of them. Logged forest is not aesthetically pleasing, and it can never be regarded as a substitute for primary forest reserves, but conservation management of exploited forests may well offer the last hope of survival for many rainforest species that require large areas of habitat.

But if logged forests are to contribute to conservation, some logging techniques will need to be modified. For example, where natural regeneration is practiced, logging operators often cut lianas and poison or ringbark non-commercial species. Many of these species are important food sources for primates and birds, and their removal is likely to reduce the capacity of the remaining forest to support them.

Despite the current power of loggers, steps can be taken to ensure that conservation interests, including the needs of local people, wildlife conservation and watershed protection, are as compatible as possible with logging. A practical strategy should be guided by three principles: establishing objectives for management, developing an adequate base of ecological information, and preparing an overall management plan.

The entire region in which logging is to take place should have an explicit—and public—management plan that specifies how its objectives are to be attained and how profits earned from timber extraction are to be reinvested to maintain the productivity of the area. Because many wildlife species cannot survive in altered conditions of microclimate and food supply, reserves of mature forest should be maintained, and even those areas allocated for timber extraction should be left with an inviolate core of primary forest—especially along ridgetops and watercourses, but also including some lowland forest. These unlogged corridors would also control erosion, serve as a natural source of seeds, and protect water quality. Finally, cleared areas should be surrounded by areas of forest, which may be logged but neither cultivated nor penetrated by modern hunters.

While many governments have taken the position that the needs of economic development dictate rapid harvests of forests to yield foreign exchange earnings, they may come to realize that in the longer term, the national interest is best served by returning much forest management to local communities.

Loren McIntyre

The Original Conservationists

People have been living in the tropical forests of Asia and Africa for hundreds of thousands of years, and in tropical America for 20,000 years, and their presence has greatly influenced both the diversity and distribution of the species in the forest.

Through the use of fire and shifting cultivation in particular, human influences on the tropical forest have been pervasive, and even the ecosystems that appear most "natural" have been significantly altered by humans at some point in the past. Professor J.E. Spencer, of the University of California at Berkeley, found that virtually all Asian forests had been cleared at one time or another

Natural dyes are one of many forest products. Here an Indian demonstrates how annatto, a widely used body paint, is extracted from the seeds of the urucu tree.

during the past 10,000 years, mostly by shifting cultivators. Today, the larger animals of mainland tropical Asia are all adapted to feeding in forest clearings and have therefore greatly benefited from traditional cultivation. In Amazonia, tribal peoples harvest certain plants and animals in ways that significantly alter their ecosystems to better provide them with the most-desired products of nature. Archeologists have found signs of past complex cultures in remote forest areas of Sumatra, Brazil, and Cambodia. Some experts conclude that environmental change and disturbance may be required to maintain a species-rich tropical landscape.

Tropical forests therefore need protection against over-exploitation, but not necessarily against people. On the contrary, traditional forest inhabitants can actually enrich the environment while preserving ecological stability and biological diversity. In Central America, indigenous people have devised ways of living in the forest that combine regular rotation of agricultural plots, harvesting natural products from the jungle, and managing wildlife. These mixed agricultural and forestry systems produce more labor and more commodity per unit of land, are more ecologically sound, and result in more equitable distribution than other practices currently being imposed on these lands.

But the forest dwellers have often been treated as opponents of conservation rather than partners. Establishing protected areas has sometimes involved

Alain Compost/Bruce Coleman Limited

removing people from forests where they have lived for many generations, thereby causing a backlash that can threaten biological resources.

Governments have only recently begun to realize the contribution that native peoples can make to conservation. During the past few years, more than 12 million hectares (30 million acres) have been granted to indigenous peoples in Colombia, and another 6 million hectares (15 million acres) are under negotiation. This land is the collective property of the Indian communities, and it cannot be sold or transferred to non-Indians. In granting these rights, the Colombian government has seen that the Indians offer the best hope for finding ways to utilize the forest without causing long-term damage.

In other parts of the tropics, relatively modest development efforts can often be far more cost-effective than large projects that bring major changes to both communities and the forest ecosystems. These range from "extractive reserves" in Brazil to eco-tourism in Panama. In Peru's Palcazu Valley, the Amuesha Indians are trying a new approach to sustainable forestry management. Using oxen, strips of forest about 20 meters (65 feet) wide are cleared. The large trees are taken to a local sawmill, and the smaller ones are pressure-treated with preservatives for use as fence posts. The benefits are earned by the Amuesha, and the clearings are both narrow enough to provide for seed dispersal from the adjacent forest and wide enough to provide sufficient sunlight for fast-growing trees. The strips are rotated over a period of several decades, effectively mimicking natural forest regeneration and providing a sustainable harvest to the Indians without seriously degrading the forest.

THE RUBBER TAPPERS OF BRAZIL

Providing local resource-users with the responsibility for using their resources sustainably can often be an incentive for conserving the larger ecosystem. An outstanding example involves the rubber tappers in the Amazonian region of Brazil, where some 500,000 people earn a living by collecting latex from wild rubber trees. The value of the forest products collected in the province of Acre in 1980 totalled US$26 million. In Acre, the per-hectare value of extraction is more than twice that of cattle ranching, and since 1970 the per-hectare value of extraction has increased more than that of either agriculture or ranching.

However, the tappers do not have title to the forests, so they have organized a series of cooperatives aimed at gaining legal guarantees for maintaining their non-timber-extractive uses of forest lands. The National Council of Rubber Tappers, founded in 1985, is therefore creating "extractive reserves"—protected areas in which forest products are sustainably harvested by indigenous communities. By granting use rights to the tappers, policy makers—with support from the World Bank and the

Inter American Development Bank—are protecting the forests against other uses. The sustainability of extraction, and the fact that it does not destroy forest, makes rubber tapping a particularly attractive alternative to agriculture and cattle ranching.

The Acre Pro-Indian Commission (CPI) has helped such communities establish cooperatives and earn recognition for Indian land rights covering nearly 15,000 square kilometers (6,000 square miles), almost 10 percent of the land area of Brazil. Given the legal guarantee to land rights that Indians hold under the Brazilian constitution, and the explicit will of the groups to preserve the mixed economy that they practice (small-scale agriculture, hunting, fishing, and rubber and brazil nut gathering for cash income), the indigenous communities of Acre are an important constituency in support of forest protection.

A New Forestry

As rainforests diminish, more intensive management will be required to maintain maximum diversity of species. The new forestry will depend upon more research:

• *Traditional knowledge* The indigenous people's traditional knowledge should be recorded, especially where the knowledge is in danger of being lost, and then applied to modern resource management. Researchers should work with indigenous co-investigators in all phases of their research design and implementation, the long-term aim being the creation of an indigenous scientific community that includes both traditional expertise and acquired scientific skills and procedures.

• *Economics* Because much over-exploitation of tropical forests has been based on faulty economics ("market failures"), we need to do more research to demonstrate the real values of tropical forests managed for multiple uses, or left in a natural state; we should assign values to non-marketed biological resources.

• *Inventory* How many species live in tropical forests, and what explains their geographical variability? If the number of species in tropical forests is 30 million, it would take 4,000 years to describe them all at the current level of effort! Quicker ways to record must be found, and research priority must be given to species that inhabit vulnerable environments.

• *Monitoring* Accurate and objective assessments of the rate of loss of tropical forests are needed, as are efforts to monitor the status and trends of specific forest areas. As protected areas become increasingly different from the surrounding lands, managers will have to monitor issues such as climate change, effects of pollution, siltation and sediment load from rivers and streams, the effects of forest edges and population dynamics in small patches, ecological relationships between species, migration, and human interactions with the forest.

Alain Compos/Bruce Coleman Limited

Sorting and checking the latex harvest. One common modern technique to coagulate raw latex for shipment is to float it on a solution of acetic acid, from which it can be skimmed off and passed through rollers to remove excess moisture. The more traditional method is to smoke it over a low fire, rolling it round a wooden stick to form a rough ball, like those being readied for shipment here.

• *Ecology* Despite relatively large investments in tropical research, many basic questions about how tropical forests function still remain unanswered. Where are the "hot spots" with particularly high levels of biodiversity? How much diversity is required to maintain different levels of ecosystem productivity and ecosystem function? How much is "redundant" in a tropical forest? In an attempt to address this concern in particular, the United Nations Educational, Scientific and Cultural Organization, UNESCO, has just launched a major international research program.

International Cooperation

The report of the World Commission on Environment and Development has given solid support to increased public concern over environmental issues. A new international convention on biodiversity is being negotiated under the leadership of the United Nations Environment Programme (UNEP), the World Bank is working with UNEP and the United Nations Development Programme (UNDP) to explore new ways to provide major funding, and a number of international conventions and programs have been set up to encourage tropical forest conservation. These are having major impacts on tropical countries from Costa Rica to Madagascar.

Opposite. *Rubber tapping in West Java, Indonesia. Though several plant species produce rubber, almost all commercial natural rubber is extracted from the tree* Hevea brasiliensis, *a native of Brazil that has been widely planted in India and Southeast Asia. Though there are a wide range of synthetic rubbers available, the natural product is still heavily in demand, especially because of its superior heat-resistant qualities.*

A baby ring-tailed lemur investigates the edible qualities of a tamarind fruit held by its mother. The lemurs of Madagascar are unique inhabitants of one of the most desperately threatened of all rainforests.

Opposite. *Tropical rainforest in a national park in Queensland, Australia. The creation of national parks and wilderness areas is essential to the future of natural rainforest ecosystems, but establishing such areas poses a formidable array of interlocking social, political, and economic problems. And these problems are not confined to the so-called Third World. Australia, the only "developed" nation possessing rainforest within its borders, has yet to succeed in establishing what most experts would consider a satisfactory level of protection for its rainforests.*

MADAGASCAR: MOBILIZING INTERNATIONAL SUPPORT

As a result of its long isolation from mainland Africa, Madagascar has many species that are unique. Twenty-eight of the island's 30 species of primates are found nowhere else. Eight of the 9 species of carnivores, 237 of the 269 reptiles, and 142 of the 144 amphibians are unique to the island, as are about 80 percent of its 7,900 plants. Yet over 90 percent of the richest habitats—the rainforests—have already been cleared, and pressures on the remainder are very high. It is clear that Madagascar must be a very high priority for conservation action.

But the conservation institutions of Madagascar are very weak, with few staff and an annual budget (exclusive of salaries) of less than US$5,000. The national economy is weak, with a high debt burden and declining per capita income. To remedy this situation, a consortium of international supporters worked with the Malagasy government to define a system of protected areas and determine the requirements for support. A budget of US$85.5 million was required to implement the Madagascar Environment Program, under the coordination of the World Bank. This was to be complemented by another US$220 million for rural development, designed to provide better living standards for people and thereby reduce their pressure on the remaining natural areas. At the time of writing, some US$67.2 million has been raised for the conservation component, mostly as grants. A national system of 77 protected areas has been established, covering 823,978 hectares (333,594 acres) or about 1.4 percent of the land area, and other areas are to be established.

INTERNATIONAL PROGRAMS IN PLACE

Some of the important projects and conventions established to support tropical forest conservation include:

- *The World Heritage Convention* The convention was established in 1972 and now has 117 member states. The Convention enables governments to nominate areas of "outstanding universal value" for the World Heritage List and also provides aid and technical cooperation to participating governments. The growing number of important tropical forest regions listed as World Heritage Sites is providing the foundation for a global network of tropical forest protected areas.
- *The International Tropical Timber Agreement (ITTA)* This commodity agreement, which has been agreed to by more than 40 nations, including countries that contain over 90 percent of the tropical forests, establishes a system of cooperation between the timber producing countries and their major consumers. It is aimed at ensuring that production is compatible with the long-term productivity of the forest. This is the first commodity agreement with conservation as an explicit objective, and its achievements to date are modest, but it provides a useful forum to discuss issues such as appropriate prices for tropical forest products.
- *The Tropical Forestry Action Plan (TFAP)* Launched in 1985 by the Food and Agriculture Organization of the United Nations' (FAO) Committee on Forest Development in the Tropics, the TFAP was designed to change investment patterns and coordinate international support for forestry. In addition, it was intended to catalyze action to develop forest resources and improve the living conditions of the rural population.Eighty-one countries participate in TFAP, each being responsible for their own forests, and it is supported by more than 40 donor agencies, which provide expert assistance in developing national plans. As a result of TFAP, political awareness of forestry issues has increased, tropical forests are higher on political agendas, budgets to forestry departments have grown, and international financial assistance to forestry has more than doubled.

However, serious problems have also arisen. Critics of TFAP have claimed that not only has it failed to improve management of tropical forests, but it has actually accelerated the already catastrophic rate of forest loss. Major investments have been made to open up logging in areas that were formerly protected by their remoteness (for example, TFAP has called for a doubling of logs from Cameroon, and a five-fold increase from Amazonian Peru). Only minor funding has been allocated to conservation. The interests of local people are being ignored, and both local and international non-governmental conservation organizations are being excluded from the process. Perhaps the greatest problem with TFAP is that it was conceived too narrowly. It deals almost exclusively with forestry institutions, ignores many other sectors with an interest in tropical forest use, and has not considered the larger issues of the use of forest-covered land.

In an effort to salvage the positive elements of the TFAP, the United Nations is considering a global convention on the conservation and development of forests. Such a convention would commit governments to specific targets for the maintenance and management of forests. With the support of wealthy governments, it could perhaps commit them to conserve sites that are of particular importance for the species and ecosystems they contain. However, little enthusiasm has been shown for this convention, and it seems that many governments prefer to see forestry issues included in conventions addressing conservation of biological diversity or global climate. Many other initiatives are being generated by local and international groups. Combined with greater political will on the part of the tropical governments and increased funding from the relatively wealthy northern countries, the transition from rapid exploitation to sustainable use is a valid hope for the coming decades. ■

BIOTOPIA

ANDREW MITCHELL

"Biotopia", the world's first permanent canopy research station.
1. Satellite dish relays data worldwide
2. Gondola access system
3. Solar panel power generation
4. Scientists working from canopy investigation pod
5. Central station—laboratory and living quarters
6. Aerial walkways for canopy access between jib cranes and laboratory
7. 80-meter (260-foot) jib crane provides access to 650,000 cubic meters (23 million cubic feet) of rainforest with sensors and canopy investigation pods
8. Airship with skyraft

In the 1970s, botanists and zoologists began to make comparisons of the plants, insects, and birds of the rainforest canopies of Panama, Papua New Guinea, and Sulawesi in Indonesia, aided by walkways and scaffolding towers built to my design, with resources supplied by the Scientific Exploration Society of Great Britain. However, the forest fringe remained beyond the reach of naturalists until Professor Francis Hallé invented his revolutionary "Radeau des Cimes", or Sky Raft.

The Sky Raft is lifted into position by a small airship filled with hot air and powered by propellers. The raft, twice the size of a tennis court, is made of kevlar netting stretched between inflatable neoprene tubing. Researchers gain access to it via ropes dropped beneath the raft to the forest floor and then use it as a platform from which to conduct experiments and collect samples before the airship returns to move it to a new site.

The canopy scientist's ultimate dream of repeatable measurements from all parts of the forest throughout the year is now coming even closer. The Smithsonian Institution has placed an experimental tower crane in the Panamanian rainforest (without damaging the trees), and researchers, suspended in a cage from the crane's jib arm, can now explore any part of the canopy, and much of the forest beneath it. The arm can be fitted with sensors that take sweeps of the forest roof to measure topography, wind patterns, and leaf growth. A larger jib crane could make 650,000 cubic meters (23 million cubic feet) of forest accessible. Plans exist to lower such a crane by helicopter at several research stations, the first at the Smithsonian's tropical research center in Panama.

Perhaps the time is not too far away when a permanent research station will be created in the forest canopy. I imagine a kind of "biotopia", where scientists could branch out from a central base to visit their research sites in gondolas suspended from towers, using walkways between platforms and even having access to a roving air-borne vehicle to gather samples from great distances. A tourist facility might be incorporated into the plans, with the revenue helping to fund research. The visual impact on a small part of the forest would be offset by the vast knowledge gained about the rainforest roof. The challenges of access and exploration are immense, but the rewards of research to agriculture, pharmacology, and to our environment—to say nothing of the human spirit—are inestimable. ●

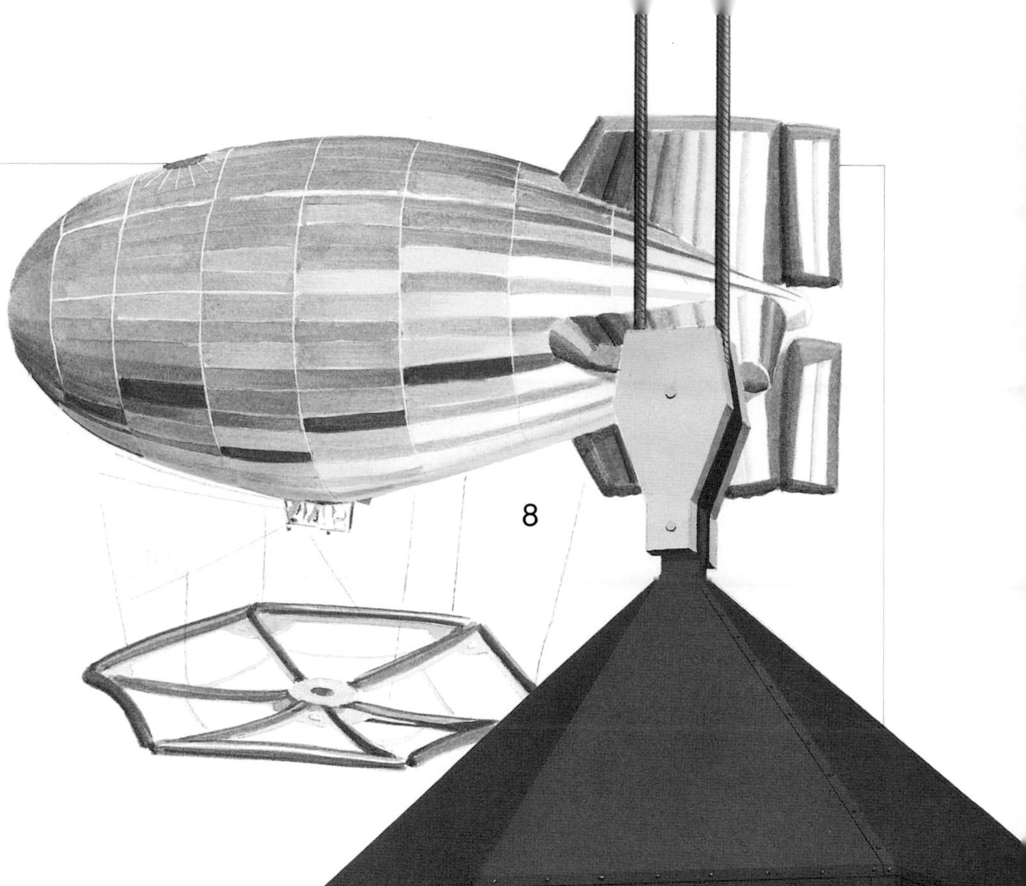

8

7

2

4

Notes on Contributors

PETER BERNHARDT
Dr. Bernhardt specializes in pollination biology and floral evolution. He currently holds two positions: Associate Professor of Botany at St. Louis University, Missouri, and Research Botanist at the Royal Botanic Garden, Sydney, Australia. He was first plunged into the tropics in his early twenties when he spent two years on a Peace Corps–Smithsonian Institution appointment as Visiting Professor of Botany at the University of El Salvador, working on endangered orchids and other epiphytes. During the late 1970s and early 1980s, Bernhardt completed his doctorate (on the pollination of *Amyema* mistletoes) at the University of Melbourne, Australia, and then worked in the University's Plant Cell Biology Research Centre. Dr. Bernhardt's current project includes a two-year field study on the pollination of guinea flowers (Hibbertia).

FRANCIS H.J. CROME
Francis Crome is a Principal Research Scientist for the Australian Commonwealth Scientific and Industrial Research Organisation's (CSIRO) Division of Wildlife and Ecology. His research has taken him into many different habitats, but he is particularly interested in the ecology and management of wetlands, rainforests, and endangered wildlife. He is currently based at the Tropical Forest Research Centre in northern Queensland, Australia, and his present research includes analysis of the effects of fragmentation on rainforest communities, methodology for assessing environmental impact, cassowary conservation, and rainforest management. He has published numerous articles on topics ranging from plankton to gamebirds, and has recently contributed a chapter on pigeons and sandgrouse to the *Encyclopedia of Birds* (1991).

JAMES A. DUKE
Dr. Duke is an Economic Botanist with the Germplasm Services Laboratory of the United States Department of Agriculture, and he is currently working on an encyclopedia of economic plants. Dr. Duke first became interested in neotropical ethnobotany in the 1960s at Washington University and the Missouri Botanical Garden in St. Louis, and it is his overriding interest to this day. For the last decade, Duke has lectured widely on the subjects of ethnobotany, herbs, medicinal plants, and new crops and their ecology. He has written numerous books and publications, including most recently *Ginseng, a Concise Handbook* (1990) and with co-author Steven Foster, a *Peterson Field Guide to Eastern/Central Medicinal Plants* (1990). Among other organizations, Duke belongs to the American Society of Pharmacognosy, the Council of Agricultural Science and Technology, the International Society for Tropical Ecology, and the Washington Academy of Sciences.

TERRY L. ERWIN
Specializing in the study of coleopterans, Dr. Erwin works in the Smithsonian Institution's Department of Entomology. His current research focuses on the natural history of carabid beetles in tropical forest canopies and the biogeography and evolution of these beetles in the Amazon Basin. First introduced to beetles at San José State College, Erwin specialized in carabid beetles during a Masters and PhD. He gained two postdoctorals, one at Harvard and the other at Lund University, Sweden, where he worked with Professors Edwards, Ball, Darlington, and Lindroth—all fellow-coleopterists. Dr. Erwin first began his tropical studies in the early 1970s in Panama and during that decade carried out extensive research throughout Central America. For the past 10 years Dr. Erwin has concentrated his studies around the Amazon, particularly in Peru.

ADRIAN FORSYTH
Adrian Forsyth is the Indonesia Program Director for Conservation International in Washington DC. Dr. Forsyth has worked in the rainforests of Central and South America and Southeast Asia for 20 years and is a prolific writer on natural history and conservation. He has written eight books on various aspects of these subjects. Dr. Forsyth is also Adjunct Associate Professor of Biology at Queen's University, Ontario, Canada.

CARL F. JORDAN
Beginning in the mid-1960s, Dr. Jordan worked for almost a decade as an ecologist with the United States Atomic Energy Commission, studying the effects of radiation on tropical forests and the fate of radioisotopes in the environment. In 1974 he joined the Ecology Department at the National Laboratory of Venezuela, where he was co-principal investigator in a multinational study on the structure and function of the Amazon rainforest, and the effects of disturbance on the ecosystem. Since 1978, when he joined the Institute of Ecology, University of Georgia, USA, as Senior Ecologist, Dr. Forsyth has been guiding graduate students in studies on the effects of forestry and agriculture on nutrient cycles in the forests of Mexico, Costa Rica, Brazil, and Thailand. Dr. Jordan's current areas of interest include tropical ecosystems, tropical forest management, nutrient cycling, and agroforestry, and on these subjects he is presently conducting projects in Brazil and Thailand.

JUSTIN KENRICK
Justin Kenrick is currently engaged in PhD research with the Department of Social Anthropology at the University of Edinburgh, Scotland. By way of fieldwork he is researching the means by which conservation and development initiatives can be directed by indigenous knowledge, motivations, and strategies. Prior to his recent anthropological studies, Kenrick worked mainly in the peace movement, and was involved in public direct action, group, and individual work.

AILA KETO
Dr. Keto is a biological scientist who has specialized in the area of rainforest conservation for the past decade. She has received several awards for her outstanding work in relation to the protection of Australia's tropical rainforests, including an inscription on the United Nations Environment Programme's "Global 500" Roll of Honour. Dr. Keto has written extensively on the value of tropical rainforests and the impacts of destroying them. She has served as an environmental consultant to the Australian and Queensland governments and has acted as a resource specialist to the United Nations Development Programme. She has a strong interest in the subjects of biodiversity and ecologically sustainable management of tropical forests, particularly with regard to timber production, and she has participated in several international conferences on these matters. Dr. Keto is currently President of the Rainforest Conservation Society, Australia.

THEODORE MacDONALD, JR.
Dr. Macdonald has been the Projects Director for Cultural Survival since 1979 and he is also a Research Associate for the Department of Anthropology, Harvard University, a position he has held since 1982. In addition to his role as Cultural Survival's Project Director, he is also Director of "Forest Residents and Forest Managers", a program for minimizing social conflict and maximizing popular participation in conservation and resource management. During the 1970s Macdonald was Anthropology Professor at the University of Illinois and the University of Wisconsin respectively. During this period he also carried out dissertation research on the socio-cultural impact of cattle raising on Amazonian Indians.

JEFFREY A. McNEELY
Jeffrey McNeely has worked for the World Conservation Union (IUCN) in Switzerland since 1980 and is currently the Chief Conservation Officer. He is responsible for conservation policy work, particularly in the areas of biological diversity, economics, and protected areas. Much of his work focuses on tropical Asia. Earlier in his career, McNeely worked in Thailand as a Peace Corps Volunteer, and there he assisted Dr. Boonsong Lekagul in writing *Mammals of Thailand*. He subsequently studied the relationship between people and nature in northeastern Nepal, assisted in the development of a system of protected areas in Laos, Cambodia, and Vietnam, and for three years ran the World Wide Fund for Nature's (WWF) Indonesian program. McNeely has written or edited a dozen books on conservation issues ranging from economics to biodiversity to culture. His major interest is in influencing government policies in favor of biological and cultural diversity, a field in which he advises the World Bank, the Asian Development Bank, a number of United Nations organizations, and governments in all parts of the world.

ANDREW MITCHELL
As Deputy Director of the international conservation organization Earthwatch Europe, Andrew Mitchell administers the provision of human and financial resources to field scientists worldwide. Mitchell spent some 15 years organizing field research programs, and during that time led pioneering studies of the rainforest

canopy using aerial walkways. He is now one of the world's leading authorities on canopy research and rainforest conservation. From 1977 to 1982 he was Scientific Coordinator of the Operation Drake expedition. In 1987 he joined the BBC and produced programs for "Tomorrow's World", "Horizon", and for the BBC's Natural History Unit. He also presented the Odyssey series for Channel 4 and was rainforest adviser for the feature film *Greystoke—The Legend of Tarzan*. He has written and presented several nature series for BBC Radio and has written seven books on the world's forests, including *The Enchanted Canopy* (1986).

NORMAN MYERS
Professor Myers is a consultant in environment and development and has worked for more than 20 years in this field, with particular emphasis on gene reservoirs and tropical forests. His main professional interest lies in resource relationships between developed and developing worlds. He has acted as consultant for many organizations including the Rockefeller Brothers Fund, the United States National Academy of Sciences, the Organization for Economic Cooperation and Development (OECD), the Smithsonian Institution, various United Nations agencies, and the World Conservation Union (IUCN). His academic and extended appointments include Chairman and Visiting Professor in International Environment at the University of Utrecht, Regents Professor and Albright Lecturer at the University of California, Research Associate at the University of Oxford's Forestry Institute, and Senior Fellow, World Wide Fund for Nature–US. He has published numerous articles, papers, and books, including the *The Sinking Ark* (1979) and *The Gaia Atlas of Planet Management* (1985). Among various forms of recognition he has been elected a Fellow of the American Association for the Advancement of Science, a Fellow of the World Academy of Art and Science, and he has been elected to the United Nations Environment Programme's "Global 500" Roll of Honour. Among numerous prestigious awards, he has received the Gold Medal and Order of the Golden Ark of World Wide Fund International, the Gold Medal of the New York Zoological Society, and the first Distinguished Achievement Award of the Society for Conservation Biology.

FRANCIS E. PUTZ
Professor Putz is Associate Professor of Botany and Forestry at the University of Florida, Gainesville. The general focus of Professor Putz's research has been the mechanisms that influence the structure and dynamics of plant communities, particularly tropical forests. This research has taken him to lowland and mangrove forests and tropical montane forests the world over. In the interests of promoting rational use of tropical forest resouces, he now dedicates much of his time to teaching and research concerned with tropical silviculture. His major goal and challenge is to help develop tropical forest management methods that are ecologically sound, economically viable, and socially sensitive.

MICHAEL H. ROBINSON
Dr. Robinson is the Director of the Smithsonian Institution's National Zoological Park in Washington DC, a position he has held since 1984. Previously, Robinson worked at the Smithsonian Tropical Research Institute in Panama, first as Biologist and ultimately as Deputy Director. His research there covered various aspects of tropical biology including predator–prey interactions, predatory behavior, anti-predator adaptations, courtship and mating behavior, phenology, species diversity, and complex symbioses. He has worked in more than 20 tropical countries and is the author of more than 100 scientific and popular articles. His publications include a book on courtship and mating behavior in spiders, and he has edited a volume on the human–animal relationship. His favorite animals are cats.

PAUL SPENCER WACHTEL
Paul Wachtel is currently the Head of Creative Services for World Wide Fund for Nature International, and he is responsible for developing international fund-raising campaigns. Wachtel has spent many years outside his native America, working for various organizations including the United States Peace Corps in Sarawak, Borneo. He also worked as creative director of several advertising agencies in Singapore and Jakarta, and as a freelance journalist. Wachtel was co-author with Jeffrey McNeely of *Soul of the Tiger* (1988) and *The Eco-Bluffer's Guide to Greenism* (1991). He is also the author of more than 100 articles on the environment, development, and unusual subjects such as medicinal leeches.

MICHAEL WILLIAMS
Dr. Williams is a Reader in Geography at Oxford University and a Fellow of Oriel College. He has lived and taught in Australia and the United States as well as in Britain, and he has long been interested in the initial settlement and landscape evolution of these countries. He is particularly interested in the processes of wetland draining and forest clearing, and is currently involved in assessing global transformations of land use and land use cover. He has edited, written, or contributed to more than 80 articles and books, including *Wetlands: A Threatened Landscape* (1991) and *Planet Management* (1992), and is currently working on a comprehensive historical geography of global deforestation. He was editor of *The Transactions of the Institute of British Geographers* from 1983 to 1987, and was elected Fellow of the British Academy and Honorary Fellow of the American Forest History Society in 1989.

ACKNOWLEDGMENTS

The publishers would like to thank the following people for their assistance in the preparation of this book:
Robert Bailey
Sue Burk
Greg Campbell, for preparation of silhouettes, pp. 16, 138
John Crawley, for development of "Biotopia" illustration, pp. 154–155
High Q Resolutions Pty Ltd, St Leonards, Sydney
Michael Kennedy
Leonn Satterthwait, Department of Anthropology, Queensland University
Mary Smith of National Geographic Magazine, Washington DC

Most of the illustrations prepared for this publication were based on original references provided by the contributors. Other sources of illustrations are listed below:

Pages 22–3 *Distribution of the world's tropical rainforests* is adapted from T*he Last Rainforests*, Mitchell Beazley, London, 1990.

Page 26 *The water cycle* is adapted from P. Raven and G.B. Johnson, *Biology*, Times/Mirror/Mosby College Publishing, St. Louis, Figure 25–2.

Page 44 *Where do they grow?* is adapted from D.R. Johansson, 1975, "Ecology of epiphytic orchids in West African rain forest", *American Orchid Society Bulletin*, 44: pages 125–136.

Page 61 *Cross-sections of the stems of lianas* is adapted from *Jungles,* Jonathan Cape Limited, London, 1980, page 141.

Pages 79 Detail of mycorrhizal fungi is based on electron micrograph appearing in "Direct phosphorus transfer from leaf litter to roots", R. Herrera, T. Merida, N. Stark, C.F. Jordan, *Naturwissenschaften* 65 (1978), page 208.

Page 121 *Cleared forest in Barbados* is based on D. Watts, 1988, *The West Indies: patterns of development, culture and environmental change since 1492*, Cambridge University Press, Cambridge, page 185.

Page 124 *Decrease in forest cover in São Paulo, Brazil. 1500–2000* is based on K. Oedekoven, 1980, "The Vanishing Forest", *Environment, Politics and Law*, 6 (4): page 186.

Pages 134–5 Rainforest regeneration is adapted from *The Gaia Atlas,* Gaia Books Limited, 1987, pages 30–31.

INDEX